South-South Development

South-South Development examines the historical background for the current situation: why it suddenly took off again approximately a decade ago; the various vectors of engagement and how they are interrelated; the actors involved; how the revitalisation of South-South development has affected development cooperation 'as it was'; and finally, how it affects the rest of the Global South.

Based on primary research on how Southern actors – via investments, aid, and trade – are changing the face of development both in the Global North and the Global South, this book contextualises the current debates, provides a systematic overview, and brings together the key themes in South-South development. It explains how countries like China, India, and Brazil are influencing domestic politics in other countries of the Global South, how they invest, and how their aid alters power structures between 'new' and 'old' donors locally. It also explains migration patterns, how they use soft power tools, and how the global governance system is changing as a result of this. This comprehensive and student-focused book includes well developed pedagogy such as text boxes, chapter summaries, key questions, bibliography, weblinks, and annotated further reading.

This book offers a unique combination of in-depth insights and secondary data on South-South development, presenting a 'state-of-the-art' account of South-South development aimed at students as well as practitioners in disciplines as diverse as International

Development Studies, International Relations, Geography, Anthropology, Global Studies, and International Political Economy.

Peter Kragelund is Head of the Department of Social Sciences and Business, Roskilde University, Denmark.

Routledge Perspectives on Development

Series Editor: Professor Tony Binns, *University of Otago*

Since it was established in 2000, the same year as the Millennium Development Goals were set by the United Nations, the Routledge Perspectives on Development series has become the pre-eminent international textbook series on key development issues. Written by leading authors in their fields, the books have been popular with academics and students working in disciplines such as anthropology, economics, geography, international relations, politics and sociology. The series has also proved to be of particular interest to those working in interdisciplinary fields, such as area studies (African, Asian and Latin American studies), development studies, environmental studies, peace and conflict studies, rural and urban studies, travel and tourism.

If you would like to submit a book proposal for the series, please contact the Series Editor, Tony Binns, on: jab@geography.otago.ac.nz

Cities and Development, 2nd edition
Sean Fox and Tom Goodfellow

Children, Youth and Development, 2nd edition
Nicola Ansell

Information and Communication Technology for Development (ICT4D)
Richard Heeks

Media and Development
Richard Vokes

Education and Development
Simon McGrath

Postcolonialism, Decoloniality and Development, 2nd edition
Cheryl McEwan

South-South Development
Peter Kragelund

For more information about this series, please visit: www.routledge.com/Routledge-Perspectives-on-Development/book-series/SE0684

South-South Development

Peter Kragelund

Routledge
Taylor & Francis Group

LONDON AND NEW YORK

First published 2019
by Routledge
2 Park Square, Milton Park, Abingdon, Oxon OX14 4RN

and by Routledge
52 Vanderbilt Avenue, New York, NY 10017

Routledge is an imprint of the Taylor & Francis Group, an informa business

British Library Cataloguing-in-Publication Data
A catalogue record for this book is available from the British Library

Library of Congress Cataloging-in-Publication Data
Names: Kragelund, Peter, author.
Title: South-South development / Peter Kragelund.
Description: Abingdon, Oxon ; New York, NY : Routledge, 2019. | Series:
Routledge perspectives on development | Includes bibliographical
references and index.
Identifiers: LCCN 2018042785| ISBN 9781138057722 (hardback : alk. paper) |
ISBN 9781138057739 (pbk. : alk. paper) | ISBN 9781315164731 (ebook)
Subjects: LCSH: Economic development—Developing countries. | Developing
countries—Foreign economic relations. | Economic
development—International cooperation.
Classification: LCC HC59.7 .K686 2019 | DDC 338.9009172/4—dc23
LC record available at https://lccn.loc.gov/2018042785

ISBN: 978-1-138-05772-2 (hbk)
ISBN: 978-1-138-05773-9 (pbk)
ISBN: 978-1-315-16473-1 (ebk)

Typeset in Times New Roman
by Apex CoVantage, LLC

Contents

Acknowledgements   *xi*

1 Introduction: the re-emergence of South-South Development   1
   The 3rd FOCAC meeting in Beijing: the wake-up call for the
   *   Global North 1*
   Aim of the book 2
   The Global South – a heterogeneous unit 4
   Classification of 'emerging' donors 7
   FOCAC: a platform for a new international order? 9
   The reaction in the rest of the South 10
   Reactions in the Global North 13
   Major sources of information on South-South
   *   Cooperation 18*
   Structure of the book 24
   Discussion questions 25
   Web pages of interest 25
   Notes 26
   Further reading 26

2 South-South Cooperation in historical perspective   28
   The end of colonialism, the Bandung Conference, and the
   *   Non-Aligned Movement 28*
   South-South economic integration 31
   Bilateral South-South Cooperation 32

G77, UNCTAD and the New International Economic Order 36
The temporary collapse of South-South Collaboration: the Cold War, the oil crisis, and the debt crisis 39
Conclusion 42
Discussion questions 43
Web pages of interest 43
Notes 43
Further reading 44

3   The resurgence of South-South Cooperation   45
Sustained economic growth in 'emerging' economies 45
Emerging growth of political influence 47
The BRIC(S): from an investment category to a political construct 49
India-Brazil-South Africa Dialogue Forum: BRICked up? 52
Regional Cooperation fora 55
   *The Bolivarian Alliance for the Peoples of Our America – Peoples'
      Trade Agreement (ALBA-TCP) 55*
*South-South Development comes full circle: the Nairobi protocol and the
   40 years anniversary of the Buenos Aires Plan of Action 58*
Conclusion 59
Discussion questions 59
Web pages of interest 59
Notes 60
Further reading 61

4   Vectors of South-South Cooperation   62
Aid: catalysing the growing engagement 62
Humanitarian assistance 68
Trade: booming, but structurally unequal 70
Investments 74
Migration 77
   *Interregional South-South migration: the case of Chinese migration to
      African countries 79*
Education: changing the hearts and minds of the next generation 83
*Global governance: using economic power to change global power
   relations 87*
Conclusion 88
Discussion questions 89
Web pages of interest 89
Notes 90
Further reading 91

5 Actors of South-South Cooperation   93
   The 'drivenness' of South-South Cooperation 93
   Government entities 94
   Private entities 99
   Two-way interaction 102
       African traders in China: 'Chocolate city' in Guangzhou 103
       *The shoe and leather industry in Ethiopia: creative destruction by
           Chinese investments? 107*
       *Learning from investing in Africa: Indian and Chinese oil companies in
           Sudan 109*
   Conclusion 111
   Discussion questions 112
   Notes 112
   Further reading 113

6 Effects of South-South Cooperation on development 'as it was'   114
   *The North's initial reaction to SSC: from fear to cooperation and
       attempted capture 114*
   *Discursive changes in the North: adopting the vocabulary of the
       South 118*
       Reactions from individual DAC members 120
   *Reactions from the IFIs: conditionality, debt sustainability, and voting
       power reform 121*
   *Dyanmics of the relationship: homogenising and differentiating
       processes 126*
   Conclusion 129
   Discussion questions 130
   Web pages of interest 131
   Note 131
   Further reading 131

7 Effects of South-South Cooperation on the rest of the Global South   133
   Studying local effects of SSC: an analytical framework 133
   Economic effects 137
   Political effects 143
       Angola: turning to China after an IMF led aid embargo 145
       *Cambodia: using development finance from the Southern partners to
           create 'balanced development' 146*
       Ethiopia: using Chinese money to develop the energy sector 147
       *Nicaragua: welcoming partners that respect the national development
           agenda 147*

Zambia: flexing muscles in relation to 'traditional' donors 148
Social effects 149
Conclusion 152
Discussion questions 153
Web pages of interest 153
Further reading 154

8 Conclusion   156
Major SSC insights 157
Future avenues of SSC research 162

References   *164*
Index   *178*

Acknowledgements

The writing of this book has taken roughly one and a half years but research on issues related to South-South Collaboration began more than ten years ago when I was fortunate to receive a generous grant from the Consultative Research Committee for Development Research, Denmark (grant number 932) that allowed me to pursue my interest in China-Zambia relations. Since then another large-scale grant, also from the Consultative Research Committee (grant number 11053-CBS), allowed me to further my understanding of resource-led development and the role of foreign direct investments in structural transformation in Africa. The generous support of the Consultative Research Committee for Development Research is gratefully acknowledged.

Due to this financial backing, I have been able to do fieldwork on issues related to South-South Collaboration repeatedly. Most of this fieldwork has been conducted in Zambia where I've had the privilege to work with Godfrey Hampwaye, who is not only a generous friend but also a great reader. I would also like to thank Hope Mwasa, who first opened the doors to Chinese companies in Zambia for me and since then has been an excellent sounding board on issues related to Zambian politics. As part of a large-scale research programme entitled *Successful African Firms and Institutional Change* I've spent lots of time doing fieldwork in Southern and Eastern Africa with Søren Jeppesen. He has been a great partner for discussion of ideas. Thanks Søren.

Many of the arguments in this book have been presented at conferences, seminars, and workshops. I'm indebted to organisers, discussants, and audiences for valuable feedback on my ideas. I would also like to thank Tobias Hagmann for valuable observations on my book proposal and Laurids S. Lauridsen for help and suggestions on sources regarding the first era of SSC. Particular thanks also goes to Emma Mawdsley for constant backing and encouragement as well as detailed comments on the entire manuscript, and Padraig Carmody and Adam Moe Fejerskov who read parts of the manuscript and provided me with insightful comments. Likewise, I would like to take this opportunity to thank the two anonymous reviewers for detailed constructive comments and criticism. Any errors remain my own.

I am also grateful to Andrew Mould at Routledge for repeatedly trying to convince me that I should write a book on South-South Collaboration and to Egle Zigaite, also at Routledge, for gently reminding me every now and again that the deadline was soon approaching.

Finally, I would like to take this opportunity to thank my family for bearing with me for spending my nights, my weekends, and what should have been our vacations, on writing this book.

❶ Introduction

The re-emergence of South-South Development

The 3rd FOCAC meeting in Beijing: the wake-up call for the Global North

In the late autumn of 2006 heads of states from 41 African countries and representatives of two dozen international organisations met with their Chinese counterparts for a three-day summit in Beijing to discuss the future of China-Africa relations. This summit, the third so-called Forum on China Africa Cooperation (FOCAC) marked the end of China's 'year of Africa' and as such it was used to mark Beijing's strategic ties with Africa. More importantly, however, it – intended or unintended – came to mark a new beginning of South-South Collaboration (SSC) as well as a break with North-dominated development.

China, on its part, used the summit to demonstrate its growing international political power, derived from more than two decades of rapid economic growth, to the outside world. Chinese leaders had spent all of 2006 building closer ties to the African continent. At the beginning of the year, then Minister of Foreign Affairs Li Zhaoxing, toured six African countries. In the months that followed both the Chinese President, Hu Jintao, and the Chinese Premier, Wen Jiabao, also visited several African countries. Meanwhile, the Chinese government issued its first 'Africa Policy', (Government of China 2006). Importantly, the summit was also perceived as an international 'pre-release' for the 2008 Summer Olympics. It was the largest international official gathering in China since 1949 and journalists, representatives, and businesspeople from all over the world were invited to witness China's ascending power. Therefore, the Chinese government had decorated large parts of the inner city with '*30 – foot-high posters of giraffes, elephants and African people in traditional dress*' (Alden 2007: 1), billboards saluted 'Amazing Africa' (Bräutigam 2009),[1] private traffic was limited in the inner city while

Plate 1.1 *Beijing prepares for FOCAC III, November 2006*

public traffic was increased, and the Chinese media was broadcasting in length from the meeting (Large 2008; Naidu and Mbazima 2008).

FOCAC, however, is much more than a 'pre-release' and a summit that takes place every three years. Rather, it is a platform to further economic, political and social engagement between China and the African continent as well as a platform to signal that an alternative exists to the 'Western' Washington Consensus. According to Taylor (2011), FOCAC rested in part on a frustration with the unequal benefits of economic globalisation as well as a disagreement with the focus on good governance and human rights pushed by the international financial institutions (IFIs) and countries in the Global North.

Aim of the book

This book is about the tectonic shifts in the in global power relations triggered by the resurgence of SSC. It examines the historical background for the current situation; why SSC suddenly took off again approximately a decade ago; the various vectors of engagement (aid, trade, investments, education, migration, and governance) and how they are interrelated; the actors involved; how the revitalisation of SSC has affected development cooperation 'as it was' – a key vector of engagement in North-South relations; and finally how SSC affects the rest of the Global South.

It is based on both primary and secondary research on how Southern actors – via investments, aid, and trade – are changing the face of development both in the Global North and the Global South. It is based on detailed studies of how for instance China, India, and Brazil are influencing domestic politics in African countries, how these countries invest, and how their aid alters power structures between 'new' and 'old' donors locally. While my own research has focused on the political economy of SSC in an African context, this book is not confined to Africa. In fact, it seeks to bring in examples from all over the Global South to highlight the contextual, spatial, and historical aspects of these relations while simultaneously pointing to the trends that cross these boundaries.

Importantly, this book questions the novelty of SSC and instead shows how the current era of SSC refers back to the previous era of SSC even if the political and economic context differs a lot. Essentially, this means that the current South-South institutions share many similarities with the institutions of the 1960s and 1970s, i.e. they seek to alter global power relations and give more voice to the countries of the Global South. However, in contrast to the SSC institutions of the past the current institutions do not pursue government intervention and protection but rather more liberalism. The book also explains the overall structures that govern each vector of SSC engagement, explains the magnitude and scope of each of these vectors, and discusses to what extent we can really talk about characteristics of, for instance, aid and trade from the Global South or whether it would be more appropriate to disaggregate the Global South into smaller units of analysis. Related hereto, the book makes an effort to show that all of these vectors are driven by specific actors, be it government actors, private sector actors, or civic sector actors. In contrast to most accounts of SSC, it shows that the lion's share of SSC activities are performed by private actors – not state actors. These include large-scale multinationals seeking new markets or resources in the Global

South as well as a multitude of small- and medium-sized firms either being pushed out of their home economy due to increasing competition and decreasing profit margins or pulled to a new host economy by new business possibilities (often created by state actors or multinational corporations).

The book does not end here. Instead, the description of the historical developments as well as the structures and actors involved in current SSC is used to analyse how SSC affects development institutions in the Global North as well as how it affects the people, firms, and institutions of the Global South. It shows that development 'as it was' led by the Global North is rapidly changing: the Global North is mimicking many of the procedures and norms of the Global South focusing more on the 'productive sector' (and less on the social sector) and more on mutual benefit. Simultaneously, the Global South is being inspired by the Global North. They now include issues of global governance and global public goods in their approaches. Likewise, they adopt mechanisms to enhance transparency and monitoring of activities. This, of course, has consequences for the peoples of the 'rest of the Global South'. The book shows how this part of the world is affected economically, socially, and politically.

The rest of this chapter is structured as follows. First, it clarifies what the Global South is and how we can classify emerging donors. Then it picks up from the opening paragraph and describes what FOCAC really is and explains how both the rest of the Global South reacted to this rejuvenation and how the Global North reacted to it. Finally, it provides insights into the growing body of literature on SSC and stresses some of the major weaknesses in it.

The Global South – a heterogeneous unit

According to the United Nations (UN), the 'Global South' describes less-developed countries primarily located in the Southern Hemisphere. In fact, the 64 countries with the highest Human Development Index (HDI) are considered part of the Global North, while the remaining 133 countries on the UN's HDI list are considered part of the Global South.

The Global South is the most recent concept describing materially less rich parts of the world. It is replacing concepts like Third World (vs. First and Second World), Developing World (vs. Industrialising or Developed World), Majority World (vs. Minority World) and Periphery (vs. Centre and Semi-Periphery). While most people would not care whether one term was used rather than another, the choice of concept has political and theoretical connotations. While both Third World and Developing (less-developed) World point towards a hierarchical world where everybody strives towards becoming like the First/Developed World and where implementation of specific policies (designed and enforced by the First/Developed World) is portrayed as being able to gradually transform a society from one category towards the other, the term 'Global South' does not have this political connotation. Likewise the 'Global South' does not point towards a body of literature that explains the underlying forces of development like for instance 'Periphery' that refers to dependency theory.

The 'Global South' concept is also replacing 'the South' that simply referred to the fact that most of the poor countries of the world were located south of latitude 30 degrees North. The difference is that 'Global' points towards some of the underlying forces, i.e. the neoliberalisation and economic integration of the world since the 1980s that has shaped how the world looks today.

The concept of the Global South, however, is no less problematic than all the concepts that it is trying to replace. It is still trying to unify countries as diverse as China, India, Brazil, Naura, and Sao Tome and Principe into one entity. Hence, it does not inform us about the differentiated processes taking place in different parts of the world – despite the integratedness of the world. Moreover, a large part of the Global South is not located in the Southern Hemisphere, and class, social, political, and economic inequalities exist all over the world. In fact, more poor people now live in middle-income countries than in poor ones (Sumner 2012). Importantly, the concept has no explanatory power. But for the link to processes of economic globalisation, it does not tell us why a particular country is poor or experiences social and political challenges. Finally, like all the other concepts it is trying to replace, it only makes sense vis-à-vis its antinomy, the Global North. In essence, then, this entails that the Global South cannot be defined *a priori*, but only in relation to something else.

Although the term Global South is not widely used in the Global South due to the influence of academia (offering BA and MA degrees in development studies) and development aid (referring explicitly to developing countries), it is widely used by politicians who seek to advance South-South Cooperation; by academics who, despite the concept's lack of explanatory power, prefer it for e.g. the developing world. The main strength of the concept is that it points to the fact that inequalities and exploitation exist all over the world, i.e. Global South denotes unequal power relations regardless of where they take place. It also hints at some of the underlying forces that create the increasing inequality although it does not explain them (Rigg 2007; Chant and McIlwaine 2009; Dirlik 2007).

For the purpose of this book, however, it is useful to further disaggregate the concept. Braveboy-Wagner (2016) suggests that countries of the Global South differ in terms of material capabilities (size and economic strength) and scope of political ambitions. These two attributes allow her to distinguish between three tiers of countries in the Global South, namely a tier that comprises countries that seek regional and global clout (such as Brazil, China, India, and South Africa); a second tier that comprises countries that have expressed/shown regional ambitions (such as Argentina, Chile, Mexico, and Venezuela), and a third tier that comprises 'small' countries that 'punch above their weight' (such as Azerbaijan, Cuba, Qatar, Senegal, and Singapore).

Both economic strength and political ambitions change over time, however. Venezuela, for instance, experienced impressive economic growth from the turn of the century to 2010 (increasing its GDP almost four times) whereupon growth rates oscillated and then collapsed and the Venezuelan economy is now contracting close to 15 per cent per year. Alongside this development, its political ambitions have altered: whereas it was a driving force in regional integration in Latin America this ambition is now softened dramatically (see also chapter 3). Stated differently, these boundaries between the tiers are by no means fixed, neither is the category 'Global South' static.

In her categorisation Braveboy-Wagner (2016) leaves out all the countries in the Global South that neither have economic clout nor political ambitions but are affected by issues of inequality, poverty, low capabilities etc. In order to take the changing nature of economic and political power into account and simultaneously include all the 'other' countries in the Global South, this book suggests distinguishing between two categories of countries in the Global South only, namely 'emerging South' and the 'rest of the Global South'. The

Table 1.1 *Characteristic of select 'emerging South' countries*

Country	*Population (2016) millions*	*GNI/Cap (2016)*	*GNI growth/capita/ year (2000–2016)*	*HDI (2015)*
Azerbaijan	9.8	4,760	8.82	0.759
Brazil	207.6	8,840	1.64	0.754
China	1,378.7	8,250	ND	0.738
Colombia	48.7	6,310	2.90	0.727
Costa Rica	4.9	10,840	ND	0.776
India	1,324.2	1,670	5.51	0.624
Indonesia	261.1	3,400	4.03[a]	0.689
Thailand	27.4	5,640	3.29	0.740
Uzbekistan	8.5	2,220	5.68	0.701

Notes: [a] 2011–2016

Source: (UNDP 2018; World Bank 2018)

former overlaps with the three tiers in Braveboy-Wagner's (2016) categorisation and thus refers to countries that seek to alter global power configurations; that have experienced rapid economic growth, since the turn of the century (although many have also experienced periods of recession); and that seek to use this newly gained economic power politically, but are not classified as Global North by the UN.

This definition differs from the numerous definitions of 'emerging' donors (see below) that solely focus on the aid aspect of these actors' engagement with the rest of the Global South. Rather, it takes into consideration the countries' level of development and their political and economic power and, thereby, it hints at their potential role in terms of South-South trade, investments and global governance (see chapter 4).

Classification of 'emerging' donors

The term emerging donors is the most widely used umbrella term for all the state development aid providers that are not members of the Development Assistance Committee (DAC) of the Organisation for Economic Co-operation and Development (OECD). It is trendy and signals a break with the past. Moreover, it hints at the economic development in the past couple of decades that has made these donors strive for political power via development cooperation. Nonetheless, as an analytical tool it is almost useless. It overlooks the fact that most of the donors in this group are in fact re-emerging on the development arena: China began its development cooperation with Egypt in 1955; India followed shortly after and Brazil has had development cooperation with Lusophone African countries since the 1970s. Hence, at most these donors emerged in the minds of researchers and political commentators. Likewise, the term fails to distinguish between different types of 'emerging' donors. In a 2008 article on 'non-DAC' donors, I tried to remedy this by distinguishing between whether or not they were members of the EU and OECD, respectively (Kragelund 2008). Thereby, I constructed four distinct non-DAC groups with varying importance for development 'as it was'. One group consisted of members of both the EU and OECD (e.g. Poland and Hungary); another of non-EU members that were members of OECD (e.g. Mexico and Turkey); a third group of EU members that were not part of the OECD (e.g. Estonia); and a fourth group that were neither members of the EU nor of the OECD (e.g. Brazil, China, India, Russia, South Africa, and Venezuela). The argument was that membership of the EU over time would tend to homogenise aid as part of it would be channelled via the European Commission. Likewise, membership of OECD would have a homogenising effect, as the OECD is a community of shared values (democracy and market economy). Thus, the group with potentially most power to change the aid system was the group of countries that were neither members of the EU nor of the OECD (Kragelund 2008). The concept of 'non-DAC donors', however, is not unproblematic either. First, it defines a group of donors by what it is not. Second, most of these actors do not perceive themselves as donors. Rather they discursively construct themselves as equal partners. Linked hereto, they do not follow the definition of aid set by the DAC (see Textbox 1.1). The latter problem was solved via the introduction of the not very idiomatic concept 'non-traditional state actors' (Kragelund 2012b) that sought to differentiate countries like China, India, and Brazil from the plethora of private actors engaged in development, see e.g. Richey and Ponte (2014), while simultaneously highlighting that these actors see aid as closely linked to other vectors of engagement such as trade and investments (see also chapter 4). However, it did not solve the problem of defining the group by what they are not.

A number of other terms have been proposed such as 'new' donors, 'non-traditional' donors, and 'Southern' donors. All of these concepts share similar challenges. They all seem to suggest that 'emerging donors' is a homogenous fixed group while the reality is that donors may change status as South Korea did when it became part of the DAC (Chun, Munyi, and Lee 2010). They all gloss over histories and they fail to take account of the internal heterogeneity in the group. Linked hereto, these terms also seem to indicate that DAC donors are very similar. This is not the case. Japan, for instance, has always perceived itself as being different from the rest of the DAC donors. Similarly, the so-called like-minded donors of Scandinavia and the Netherlands have stood out in terms of their insistence on poverty focus and relatively high levels of aid compared to the size of their economies (Lancaster 2008).

Despite all the problems related to the concept emerging donors it is by far the most used term to describe the state actors that either revived their development assistance programmes during the past decade or set up totally new ones. Characteristic for most of these actors is that they play a double role of being both donors and recipients (cf. Mawdsley (2012b)).

FOCAC: a platform for a new international order?

The first FOCAC meeting in 2000 was some years underway. The idea was launched by former Chinese President Jiang Zemin during his tour to six African countries in 1996. China wanted to re-establish Sino-African relations built on consensus, friendship, and mutual benefit and the Organisation for African Unity agreed that this was indeed a good idea. After some years of planning and fine-tuning, representatives from 45 African states and several international organisations attended the three-day meeting in Beijing in October 2000. The meeting was split up into four tracks, namely trade, economic reform, poverty eradication and sustainable development, and cooperation in education, science technology, and health care. These tracks came to characterise the debate in later FOCAC meetings as well. More importantly, however, FOCAC signalled a break with the international order at the time. Thus, the official idea was that China and African countries could collaborate to create a new mutually beneficial international and political order to:

> *strengthen solidarity and promote South-South Cooperation;
> . . . enhance dialogue and improve North-South relations; take
> part in international affairs on the basis of equality and in an
> enterprising spirit; . . .* [and] *look forward into the future and
> establish a new long-term stable partnership of equality and
> mutual benefit.*
>
> (Taylor 2011: 39f).

An integral part of the FOCAC is the declaration that follows each
meeting. The first declaration included a pledge from China, among
others, to cancel debts worth 10 billion Yuan to debt-ridden African
countries (fulfilled before time); increase 'foreign assistance' to
Africa; and use South-South Collaboration as a vehicle to further
economic and social development. The collaborating partners were
very pleased with the outcome of the first meeting and in the years
that followed immediately after it the number and intensity of high-
level visits between China and African countries increased. Despite
this, hardly anyone outside China and the African continent noticed
the meeting and the new phase of South-South Cooperation that it
came to signal.

The same goes for the second FOCAC that was held in Addis Ababa
in December 2003. This is in spite of the numerous concrete proposals
that were put forth in the Addis Ababa Action Plan that lay bare the
main issues confronting the partners and revealed how to overcome
them. Among the most noteworthy issues in the Addis Ababa Action
Plan were the agreements to make available tariff-exemption treatment
to some products to open the Chinese market for the least developed
countries in Africa; to strengthen agricultural cooperation; to develop
infrastructure; and to promote investment (Taylor 2011).

Hence, it was not until after the third FOCAC meeting in November
2006 where more than 1,700 delegates discussed future China-Africa
relations and where some 1,000 reporters spread the word that once
again China was a significant development partner for other Southern
countries that the world reacted to the new situation.

The reaction in the rest of the South

India responded to China's diplomatic show off by orchestrating its
2008 India-Africa summit. While by no means as spectacular as the

2006 FOCAC meeting it consisted of the same ingredients and had the same purpose. Hence, it was attended by heads of states of 14 African countries and representatives of a number of regional associations. It aimed at displaying the long-term collaboration between India and Africa and demonstrated the growing international power of India thereby changing the world's perception of India from being a poor recipient of development aid to being an emerging economy able to afford its own development aid programme[2]. The 2008 summit led to a 'Delhi Declaration' and a Cooperation Forum, and like its Chinese counterpart, the 2008 India-Africa summit was completed by the announcement of various measures including concessionary credits and duty-free access to a large number of product lines. Again, like the Chinese version, another India-Africa summit was held three years later, in 2011, while the third one was held in October 2015, this time attended by 34 African heads of state.

While there is no doubt that the 2008 India-Africa summit can be perceived as a continuation of five decades of close South-South Cooperation between India and (some) African countries, a more accurate reading of the event would factor in India's long-term competition with China for global recognition. In fact, in the decade that preceded the summit India closed down several diplomatic missions in Africa (and Asia). It was only after China rejuvenated its Africa relations that India did the same. India's renewed South-South engagement thus relates closely to its ambition to achieve global recognition including a permanent seat in the UN Security Council and to catch-up with China (Cheru and Obi 2011; Kragelund 2011; Taylor 2012; Mawdsley and McCann 2011).

Global political ambitions, a seat on the UN Security Council, and access to new markets have also played a pivotal role in Turkey's recent economic and political engagement with actors on the African continent. In the words of Özkan and Akgün (2010: 529) the incumbent government in Turkey reasons that '*if Turkey is to be a global player, it can no longer overlook a rising Africa*'. This realisation did not come overnight. During the late 1990s, Turkey's trade relations with Africa grew incrementally and due to its geographical location, between Asia and Europe, Turkey also played a central role in African politics. The Turkish government therefore in 1998 developed the 'Opening up to Africa policy' which sought to improve diplomatic (opening new embassies), political (high-level visits), economic (trade and investment protection plus technical assistance) and cultural cooperation (education) with the continent. The response from Africa was instant. Turkey was coined a strategic partner for the African Union (AU) in 2005, and President Erdoğan was invited to the opening session of the 2007 AU summit. Turkey on its side coined 2005 the year of Africa, developed an Africa Initiative, and in 2008, Turkey held its own summit, the Turkey-Africa summit coined 'Solidarity and partnership for a common future'. Like China and India, Turkey also chose to conclude the summit with a declaration as well as a cooperation framework (Özkan and Akgün 2010; Apaydin 2012). Six years later, in 2014, the second Turkish-Africa summit was held in Equatorial Guinea.

In the second half of the first decade of the new millennium, other Southern actors also rejuvenated their economic and political engagement with other Southern countries. Already in 2005, the League of Arab States and countries of South America joined forces in the Summit of South American-Arab Countries (APSA) held in Brasilia, Brazil (repeated in 2009, 2012, and 2015 in Qatar, Peru, and Saudi Arabia, respectively). Likewise, Asia and Africa rejuvenated old ties during the 2005 Asia-Africa summit, and the Arab world joined forces with African leaders in the second Afro-Arab summit in Sirte, Libya, in 2010 (the first one was held in 1977 in Cairo, Egypt) to further cooperation and political coordination. Not surprisingly, they also adopted both an action plan (plan of Joint Arab-African action) and a declaration (the Sirte Declaration). Africa and South America

Figure 1.1 *Cross-Regional South-South Summits, 2000–2015*

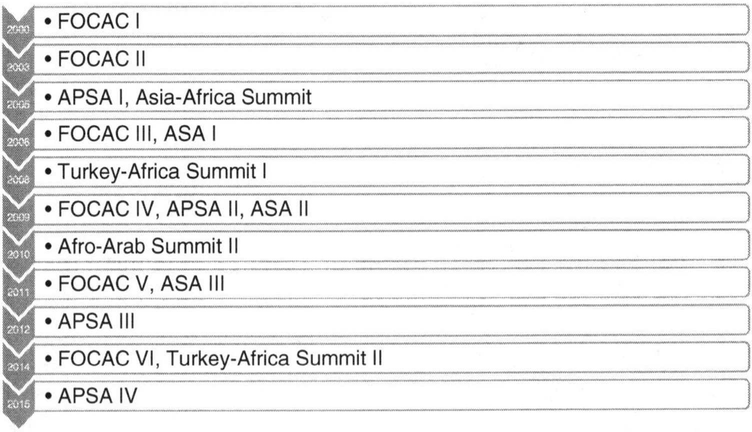

Year	Summits
2000	• FOCAC I
2003	• FOCAC II
2005	• APSA I, Asia-Africa Summit
2006	• FOCAC III, ASA I
2008	• Turkey-Africa Summit I
2009	• FOCAC IV, APSA II, ASA II
2010	• Afro-Arab Summit II
2011	• FOCAC V, ASA III
2012	• APSA III
2014	• FOCAC VI, Turkey-Africa Summit II
2015	• APSA IV

did not want to be left behind. Hence, they held their first meeting in 2006 in Nigeria, whereupon the following ones were held in Venezuela (2009) and Equatorial Guinea (2011)[3].

Development finance, either in the form of loans, grants, stipends, or debt cancellation, was often the vehicle through which Southern actors rejuvenated their engagement, which also consisted of trade, investments, migration, governance, and military support (Kragelund 2008; Mawdsley 2012b; Chaturvedi, Fues, and Sidiropoulos 2012). The re-emergence of South-South Development Cooperation, then, on the one hand is built on 'traditional' development cooperation (see Textbox 1.1) and on the other hand, spans a variety of economic, political, and cultural vectors. These vectors have the potential of transforming development paths for other Southern countries directly as they, for instance, increase the financial resources available for development, lay bare alternative strategies for achieving development, and increase competition among donors. The rejuvenation of SSC may also alter development paths indirectly as 'traditional' development actors have reacted to the new situation. They, to an increasing degree, mix the vectors of engagement – like the Southern development actors – in order to create a 'win-win' situation where both the donor and the recipient benefit from the engagement. They have also begun a process of rethinking central development concepts in order to cater for the new situation, cf. Hynes and Scott (2013).

Reactions in the Global North

Actors in the Global North also responded to the rise of the Global South in development. The first reactions came after the publication of China's Africa strategy and were repeated immediately after the third FOCAC meeting. In the words of the The Economist (2006):

The summit in Beijing [FOCAC III] *is being greeted by Chinese officials and the country's state-run media with an effusion reminiscent of the cold-war era when China cosied up to African countries as a way of demonstrating solidarity against (Western) colonialism.*

The Economist was not alone. In the years that followed several commentators voiced their concern for the rejuvenation of 'emerging' donors. This included Moises Naím (2007) who in *Foreign Policy* warned about *rogue donors* who supplied *toxic aid* and pushed *toxic ideas* as well as the former chair of DAC, Richard Manning (2006), who in a balanced account, warned that rejuvenation of 'emerging' donors may cause postponement of necessary reforms in the Global South (due to absence of conditionalities) and worsen developing countries' debt burden (due to bad terms).

What these commentators feared most was that the rejuvenation of South-South Cooperation would revolutionise development cooperation 'as it was', i.e. that it would put an end to development as a North-South endeavour led by the Bretton Woods institutions and supported by bilateral donors from the Global North. However, the reaction was in no way only one-sided. Cheru (2016) identifies four different types of reactions to the rejuvenation of South-South Cooperation, namely the 'alarmists' made up of people like Naím who ultimately perceive the phenomenon as a security concern for the US; the 'sceptics', i.e. aid bureaucrats fearing that advances in aid effectiveness will be revoked; the 'critics of new imperialism', who compare China and India's engagement on the African continent with imperialism, that is, extension of power to acquire territories; and the 'pragmatic cheer leaders', who perceive the phenomenon as an opportunity for Southern countries to augment agency as well as a chance to rethink global governance.

Notwithstanding these opposing reactions, it is important to give a very short overview of what development – and development cooperation – is in order to further our understanding of what is at stake – and why the revival of South-South Cooperation creates so much discussion. Development 'as it was' dates back to the establishment in 1944 of the two Bretton Woods institutions, the International Bank for Reconstruction and Development (the World Bank) and the International Monetary Fund (IMF), that were to govern the world economy in the aftermath of the war. Immediately afterwards the United Nations Charter (1945) was drawn up that led to the founding of several UN bodies that came to characterise development in the post-World War II era.

The official birth of international development cooperation and the initiation of the 'development age' came a few years later with the publication of President Truman's 'Point Four Programme' in January

1949 that among others things proposed a 'fair deal' for the world, to be achieved by deploying technical and financial assistance to help 'underdeveloped' countries to become 'developed'. Thereby, the Point Four Programme '*evoked not only the idea of change in the direction of a final state but, above all, the possibility of bringing about such change*' (Rist 2008: 73).

In the years that followed former colonial powers were establishing bilateral development assistance programmes. International development cooperation was slowly maturing. To help the state actors learn from each other, the Development Assistance Group was formed in 1960. In 1961, it was transformed into DAC. From being an experience-sharing entity, it changed into a norm-making entity. In particular, the DAC began a process of coordinating aid in order to improve standards of living for the inhabitants of less-developed countries. Moreover, the DAC worked to increase the amount of money available to finance development, and focused on long-term development issues (see also Textbox 1.1).

With time, the DAC became *the* hegemonic power in international development cooperation. The DAC defined the purpose of aid (to promote economic and social development in developing countries), its terms (concessional), and its rationale (for developing countries to engage in the global economy and for people to overcome poverty). In order to encourage members to abide by these aims and rationales, the DAC developed common objectives and guidelines (e.g. untying aid and spending 0.7 per cent of GNI on aid), established standards for monitoring and evaluation, and instituted recurrent peer aid reviews and high-level meetings.

For a very long time, DAC donors acted side-by-side with non-DAC donors. Thus, from the 1960s until the end of the Cold War, China, the Soviet Union, Cuba, and a number of OPEC countries provided substantial amounts of aid to developing countries. During the 1980s – and in particular with the end of the Cold War – developing countries lost their geopolitical importance; former Eastern Bloc countries competed for (diminishing) aid money; and non-DAC donors focused on domestic restructuring. Thus, DAC members came to have hegemonic power over what development is all about and how to achieve it (in most of the Global South). In the years that followed, this entailed economic liberalisation (kick-started by the debt crisis

Box 1.1 Development assistance

One of the major controversies in the debate over the effects of the rejuvenation of SSC has been the definition of aid. According to Riddell (2007) 'foreign aid' includes development assistance that seeks to meet long-term development and poverty needs as well as (short-term) humanitarian assistance, and assistance that primarily meets political/strategic needs such as military aid.

This is in contrast to what is normally referred to as 'development aid', i.e. overseas development assistance (ODA), that is defined both by its *purpose* and its *terms*. As regards the former, ODA primarily has to promote economic and social development in developing countries. The latter refers to the official and concessional character (below market interest rates) of the transfers, that is, it has to be provided by official agencies (state and local governments) and include a grant element of at least 25 per cent.

The Southern development actors abide to neither of the two definitions. In China's statistical yearbook, development assistance comprises both foreign aid and government sponsored investment activities (both at home and abroad). China's development assistance is made up of grants and interest-free loans funded by state finances (such as via the Ministry of Commerce and the Ministry of Foreign Affairs and supported technically by state-owned enterprises) and concessional loans funded by the Export Import Bank of China. Not all of the concessional loans have a grant element of at least 25 per cent and hence are not defined as ODA by DAC.

What comprises India's development assistance is no less opaque. India has no formal definition of aid, but what is normally comprised in India's accounts of its development interaction with other countries in the Global South include ODA-like flows (scholarships, small grants, and debt relief) as well as export credits and investment-like flows.

The problem is that many commentators tend to forget the difference between Southern and Northern development actors. In the words of Bräutigam (2011: 753): '*Many who comment on China's economic co-operation activities and who compare them with ODA are not themselves familiar with the framework of rules and norms governing ODA, export credits and other forms of official finance. As a result, they end up comparing apples and oranges.*'

In order to further our understanding it is thus of the utmost importance to clarify the most important concepts re. development finance. ODA is probably the most important concept. As stated above, it is defined by both purpose and terms. It excludes export subsidies and credit lines to further trade and investment (only) as well as military aid. Similarly, it excludes transfers by private actors and transfers not given on concessional terms. This sounds easy in theory. In practice, however, the degree of concessionality – or rather whether or not a loan is 'concessional in character' – is hard to determine as it is determined by the size of the donor subsidy which, among other factors, is influenced by the difference between the interest rate of the donor loan and the market rate (Bräutigam 2011).

Similarly, what may be perceived as formal rules governing ODA most often are soft agreements open to interpretations and reinterpretations (Paulo and Reisen 2010). One such example is the link between aid and trade. Since the formulation of the 'Helsinki Package of Tied Aid Disciplines' in 1991 (agreed in 1992) DAC donors have agreed to separate exports credits from ODA and part so-called 'tied aid' from 'untied aid' to make sure that trade is not distorted by aid to create comparative advantages for donor country exporters vis-à-vis other donors as well as recipient country producers.

and the subsequent structural adjustment and stabilisation programmes in the 1980s) as well as political liberalisation. It also entailed donors controlling an increasing share of policy areas in developing countries – despite a donor rhetoric revolving around partnership and ownership. In particular, broad-based schemes to lower developing countries' debt burden such as the Highly Indebted Poor Countries Initiative (HIPC) meant that donors came to dictate the majority of social and economic policies in debt-ridden developing countries (Fraser 2008a).

Thus by 2006, when the third FOCAC meeting took place, DAC donors had had more than 15 years to define what development was all about, how to achieve it, and decide who should take the lead in this process. Unsurprisingly, the 'established' donor community voiced its misgivings about the new situation. As hinted at above the reaction was varied. The immediate response was often one of fear and resentment. The rejuvenation of SSC was often perceived as bad for development as it did not link flows of development finance to democracy, transparency, accountability, human rights etc. In other words, it totally ignored 'traditional' development actors' 'good governance' agendas. In particular, the widespread use of economic and political conditionalities is being challenged by 'emerging' donors who prefer to provide aid with 'no strings attached'. Seen from this perspective the way forward was to convince 'emerging' actors to adopt DAC principles and guidelines by way of a mix of conferences, outreach programmes, and study groups. This would allow them to *'learn from and adopt Western "best practice" and superior experience'* (Mawdsley 2017a: 110). Since then the approach towards emerging donors has changed. When DAC donors realised that they were not able to persuade the big emerging donors to play by DAC rules, they themselves became more open about the purposes of aid and increasingly explicitly claimed that their aid practices revolved around 'mutual benefit'. After more than a decade since the heads of states of 41 African countries gathered in Beijing, the reading of emerging donors is much more complex. Some see the 'new' donors as providers of new ideas and experience that are readily applicable in the Global South while others work hard to minimise their economic and political power (Mawdsley 2017a; Mawdsley, Savage, and Kim 2014; Kragelund 2015).

Major sources of information on South-South Cooperation

With the publication of China's 'Africa strategy' in January 2006 and the widely broadcast FOCAC meeting in Beijing in November 2006, the interest in SSC grew rapidly. Researchers, journalists, and policy-makers alike wanted to further their understanding of what was happening and hence the number of 'China-Africa' conferences was mushrooming. Participants in these conferences not only wanted to understand, they also wanted to draw hard conclusions about the consequences for development in general and the Global South in particular. The problem was, as Bräutigam (2009: 3) puts it in her myth-killing book about Sino-Africa relations, they were '*drawing conclusions with only scant information*'.

Luckily, many researchers took up the challenge and began prising the 'emerging' actors box open by not only studying aggregate flows of trade, aid, and investments between the rejuvenated actors and the rest of the Global South (see e.g. Kaplinsky and Messner (2008)) but also began conducting fieldwork among for instance Chinese traders in Namibia (Dobler 2008a); examining Indian actors in East Africa (Mawdsley 2010); examining donor reactions to 'new' players in Nicaragua (Walshe Roussel 2013); scrutinising Chinese and Indian oil companies in Sudan (Patey 2014); and exploring African students pursuing higher education in China (Haugen 2013).

In fact, several disciplines have taken up the subject of SSC not only providing different perspectives on SSC but also scrutinising the developments using different methods. Economists tend to probe into larger trends concerning for instance development aid and local corruption, the link between aid, trade, and investments, and the link between aid and authoritarian regimes. They use available data from for instance aiddata.org (see p. 21) and Afrobarometer surveys to further our understanding of the overall correlations between the rejuvenation of SSC and political and economic developments in the 'rest of the Global South' (Broich 2017; Dreher, Nunnenkamp, and Thiele 2011; Isaksson and Kotsadam 2018a; Biggeri and Sanfilippo 2009). While these studies have greatly improved our knowledge of the links between different vectors of engagement (see chapter 4) at the aggregate level, they do not inform us about why this is so and how it affects societies in home and host economies.

To this end, anthropologists, geographers, scholars with an international development studies background, political scientists, and sociologists have greatly improved our knowledge. Anthropologists have taken us to the local communities and cracked open the black box of how the meeting between peoples of the Global South affects day-to-day activities in the host economies (Dobler 2008a; Ndjio 2017; Pedersen and Nielsen 2013); geographers have helped us to critically engage with key SSC concepts and directed our attention to the 'spaces'/enclaves that develop as a result of the renewed South-South engagement (Carmody 2009; Mawdsley 2012a; Brooks 2010); scholars from development studies have paid particular attention to how this new trend affects development cooperation as it is (Quadir 2013); political scientists have scrutinised broader questions of power relations, e.g. how does the renewed interest in Africa affect its position in international relations (Taylor 2014; Harman and Brown 2013); and sociologists have among others analysed how SSC affects labour relations (Lee 2009).

The various actors of SSC and the vectors of their engagement have not received an equal amount of scholarly (and political) interest. By far the most interest has been devoted to Sino-Africa relations. This is also apparent in the number of books published that include the term 'China-Africa'. Figure 1.2 depicts share of books in 'Goggle books' that include the terms 'South-South Cooperation' and 'China-Africa'. While SSC has been of interest to researchers since the late 1970s, China-Africa relations only became interesting by the turn of the century and interest in this particular relationship is now outgrowing the interest in SSC more broadly.

These aggregate figures, however, do not inform us about the kind of information available. An idea of who, for instance, is writing about China-Africa relations and thus an indication of the academic 'quality' of this information comes forth from an analysis of the discrepancy between the number of 'Scholar google' and 'Social Science Citation Index' publications that include for instance both 'China' and 'Africa'

Figure 1.2   *Occurrence of 'South-South Cooperation' and 'China-Africa' in Google Books, 1975–2008*[4]

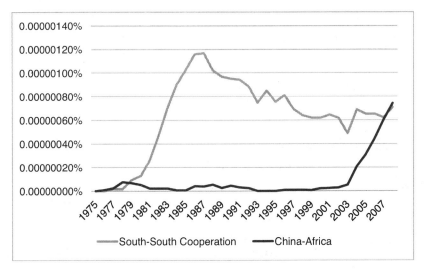

Source: https://books.google.com/ngrams/

Figure 1.3   *Number of yearly publications with 'China' and 'Africa' in title, 2004–2015*

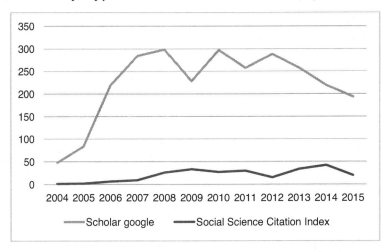

Source: Scholar.google.com and Web of Science

in the title in the past decade (see Figure 1.3). This discrepancy indicates that a fair share of the available information stems from so-called grey literature – most often not based on (long-term) fieldwork, and commonly written from a particular viewpoint or with a particular aim in mind.

Added to the politicised nature and the lack of deep understanding of the subject under scrutiny is the fact that data on South-South relationships tend to be hard to get hold of; of low quality; and hard to compare to North-North or North-South flows. Many Southern actors are not members of the OECD and hence data on, for instance, aid flows have not been collected by the DAC. This has led several researchers and think tanks to come up with various methods to determine the scope and magnitude of these relations. The challenge is that most Southern actors use definitions of aid for instance that differ from DAC. Hence, whatever is gathered is not comparable to the existing statistics on North-South flows.

Our knowledge of these flows, however, is constantly improving and databases do exist that may help us further this understanding even more (even though they are not at all flawless as will be demonstrated below). One of these databases is AidData, which is a relatively new database of development finance made up of a mix of data from OECD statistics, annual reports from donors, project documents, donor websites, and the like. In contrast to other databases, AidData deliberately adopts a definition of aid that is broader and more inclusive than ODA (see Textbox 1.1). Thereby, it caters better for the characteristics of SSC. Thus on top of what is included in ODA, AidData includes loans at market rates as long as they come from governments or intergovernmental organisations and as long as their main purpose is to bring about social and economic development among the recipients. However, the database is not comprehensive: it excludes finances from Foundations (of big importance in the Gulf states, for instance, and of growing importance in the Global North (Fejerskov 2015), as well as military aid. Moreover, the early versions of AidData lacked data for both China and India (Dreher, Nunnenkamp, and Thiele 2011). More recent versions, however, include both: data on India's development finance is based on donor documents and data on Chinese development finance (to Africa only) is based on information from media reports, scholarly articles, and official websites.

The accuracy of AidData's information on Chinese development finance has spurred a heated debate. Deborah Bräutigam, an expert in the field, reacted promptly to the first release of Chinese aid data. According to her, reliance (only) on media reports easily overestimates the scope and magnitude of Chinese development finance (Bräutigam 2013). The challenge is that the media tends to report Chinese aid three times: when the deal has been sealed, when work begins, and when the project has been finalised. In other words, there is both a risk that reported aid transfers never materialise and that the same transfer is counted three times. The first datasets on Chinese aid from AidData included projects that were only on the drawing board. In October 2017, AidData released a refined dataset on Chinese global development finance. While the methods and thus the problems remain the same the dataset is constantly improved by new information. Moreover, it classifies projects according to whether they are completed, cancelled, approved, pledged etc. As it is based on 'open source' data it is also constantly improved. One way of improving AidData is the so-called 'ground-truthing' of the flows, i.e. site-visits with recipient government officials. When Muchapondwa et al. (2016) did this in Uganda and South Africa they on the one hand found that what was on the ground reflected the descriptions in the database. On the other hand, they were only able to track down one-third of the described projects.

Another way to improve the database is by comparing it with other databases. Deborah Bräutigam and her team at the Johns Hopkins SAIS China Africa Research Initiative (CARI) adopt the time-consuming 'forensic internet sleuthing method' to assess the size of China's loans to Africa. In short, trained researchers also speaking Mandarin triangulate reports of loans, information from Chinese contractors, and information from personal contacts in China and in African countries with interviews with people on the ground. As can be seen from Table 1.2, the discrepancy between AidData (ver 1.0) information on Chinese loans and CARI data sometimes is enormous.

Table 1.2 *Select Chinese 'megadeals' in Africa according to Aiddata and CARI*

Country	Year	Project	Aiddata (USD mn)	CARI (USD mn)
Sudan	2003	Merowe hydropower	836	608
Angola	2004	Infrastructure	1,507	2,000
Zimbabwe	2004	Kariba hydropower	1,010	0
Nigeria	2006	Abuja Light Rail	673	500
Ethiopia	2009	Dams	2,249	810
Zambia	2010	Kafue Gorge hydropower	930	0

Source: Adapted from: Bräutigam and Hwang (2016)

Like aid data, data on South-South investments are inaccurate. Several multilateral agencies such as UNCTAD, the World Bank, IMF and OECD provide data on Foreign Direct Investment (FDI) flows and FDI stock. However, data from these agencies are not easy to compare when we want to investigate changes in South-South investments. First, they do not apply the same definition of 'South'; second, they do not use the same sources to estimate the size of inwards and outward investments; third, round-tripping of investments, i.e. money from one country going to another country (most often a tax-haven) and then returning to the country of origin as investments, overestimates FDI flows; fourth, the use of offshore financial centres, i.e. countries or jurisdictions that provide financial services, makes accurate estimates increasingly difficult (Aykut and Ratha 2004); and finally, data is provided at a very aggregate level. Sector-and firm-level data is available from national investment agencies in countries receiving FDI. However, like the aid data described above, local investment data is also inaccurate: national investment agencies report pledges – not actual investments – essentially meaning that many investments never materialise; and national investment agencies in the Global South seldom have the capacity to monitor the investments resulting in an overrepresentation of investments categorised in sectors characterised by massive incentives – and underrepresentation of investments not deemed important by the host economy (Kragelund 2009a).

Unsurprisingly, trade data is also often inconclusive. Comparisons of imports from country A to country B hardly ever match corresponding export data from country A to country B. Most trade analyses rely on the Harmonized Commodity Description and Coding System – in short the HS Code – that classifies every product into specific categories (see e.g. the UN Comtrade database). Hence, in principle the two figures should be the same. In practice, however, different ways of valuating goods entail that import and export figures seldom match. Likewise, time matters for comparisons: goods exported one year may not arrive before the year after. Time also influences exchange rates and hence the registered value of a good. Equally important, not all goods are shipped directly from the exporter to the importer. Often a middle-man adds extra costs along the journey, making the value of imports higher than the value of exports, and finally the exporter may not know the final destination of a good. Moreover, trade increasingly takes place between firms (as intermediate product) – not countries, but most trade databases register trade of final products between countries. Thereby there is a real risk of double counting as products cross borders several times before they reach the stage of final product and become registered in trade databases (Horner and Nadvi Forthcoming).

To the extent possible, it is therefore recommended that data is triangulated. The AidData database, for instance, is an excellent starting point for primary research. The database provides geodata on for instance Chinese aid projects in Africa, including links to additional sources of information. Likewise, data from national investment centres may provide some useful insights into which investors are most important and in what sector. However, disaggregated data is by no means accurate and has to be informed by surveys, archival studies, or interviews with investors. Finally, ethnographic studies of South-South trade has revealed how goods move across boundaries, how traders make use of differences in tariffs and how a plethora of goods that cross borders are not registered in official databases (Dobler 2008a).

Finally, it is worth noting that the literature published in English may exaggerate the importance of some actors (China and India for instance) over others (Venezuela and Saudi Arabia) and some vectors (aid and trade) over others (regionalisation). As Muhr (2016) reminds us, the Anglo-Saxon literature is to some extent blind to the developments that we see in Latin America for instance (see also Textbox 3.2); it also tends to be blind to the politicised nature of many of the initiatives that emerge in particular in Latin America: SSC in most of Latin America today is not only about technical transfers. Rather, it seeks to rejuvenate the ideas that formed the New International Economic Order (see chapter 2). Seen from this perspective, Latin American SSC is more inspired by structuralist economics than the neoliberal doctrines of the West (see Textbox 2.2). While China is indeed the most powerful current SSC-actor, it is not the only one, cf. van der Merwe (2016). Likewise, while most SSC vectors include one or more African countries, many SSC flows bypass the African continent altogether. It is thus important to bear in mind that the Anglo-Saxon literature on SSC may not reflect the SSC actors and vectors one-to-one. This literature, like many others, is also geopolitically motivated.

Structure of the book

The book proceeds as follows: chapter two present SSC in a historical context. It portrays the boom and bust of the first phase of SCC via the beginning in Bandung in 1955, the rise through the establishment G77; the call for a New International Economic Order; and the signing of the Buenos Aires Plan of Action and bust caused by rising commodity prices, intensification of the Cold War, and growth of regional power blocks. Chapter three describes the rejuvenation of SSC beginning with the economic growth in East Asia and taking off with the rise of China and India. It explains how their economic power was transformed into political power and how this gave rise to a new phase of SSC. Chapter four takes us beyond the historical accounts of SCC and seeks to disaggregate the various vectors of engagement. In doing so, it pays particular attention to aid, trade, investments, and migration but it also touches upon education and global governance. Chapter five complements this picture by presenting the main actors of SCC and explaining what drives them. Chapters six and seven seek to answer the question: what are the likely implications of the resurgence of SSC? While chapter six scrutinises the effects on other actors in the development arena, chapter seven takes us to the (rest of) the Global South. Chapter eight concludes the book by answering the following question: how does the resurgence of South-South Cooperation affect development in the Global South?

Discussion questions

- Discuss how different terms affect our understanding of the actors in the 'Global South'.
- Describe how actors in the Global South and in the Global North, respectively, reacted to the rejuvenation of SSC following the 3rd FOCAC meeting in Beijing in 2006.
- Discuss the quality of South-South information and consider strategies to overcome these deficiencies.

Web pages of interest

- The China Africa Research Initiative Blog is the official blog of the China Africa Research Initiative hosted by Johns Hopkins University. It offers regular comments/corrections on news on China-Africa relations as well as ideas for future research: www.chinaafricarealstory.com
- Forum on China-Africa Cooperation is the official FOCAC website. It contains all official documents from the meetings: www.fmprc.gov.cn/zflt/eng

- Global South Studies Center is a historically informed interdisciplinary research centre located in Cologne, Germany, that seeks to further our understanding of the dynamics of the Global South: gssc.uni-koeln.de
- South-South Information Gateway is a Malaysian-based website offering news and analyses of relevance for the Global South: www.ssig.gov.my
- United Nations Office for South-South Cooperation seeks to mainstream South-South Cooperation across the UN system and helps member states to engage in SSC: www.unsouthsouth.org

Notes

1 The Chinese decoration was not uncontroversial as it was said to use only 'typical' images of Africa.

2 The countries of the Global South use a variety of different terms to describe their development co-operation, including technical assistance, development assistance, and technical cooperation. For the sake of simplicity, these terms are used interchangeably in this book.

3 Not all initiatives that resemble South-South collaboration originate from the South. In 2007, for instance, the first Arab-Africa Initiative conference was held in Cairo, Egypt. This conference like its many successors, however, was facilitated by the UN in order to implement the Millennium Development Goals (MDG) in African countries and the Arab world.

4 Ngrams displays the percentage of books published in English in year X that contains a particular concept. As Ngrams is case sensitive, Figure 1.1 shows aggregated data for South-South Cooperation, South-South cooperation, south-south cooperation, and SOUTH-SOUTH COOPERATION.

Further reading

Bergamaschi, I., P. Moore, and A. B. Tickner. 2017. *South-South Cooperation Beyond the Myths: Rising Donors, New Aid Practices?*: London, Palgrave Macmillan. This edited volume traces the ideas, identities and actors of ten different Southern donors, including, but not limited to China, India, and Brazil. Unlike most other books on this issue it also includes chapters on Colombia, Cuba, Venezuela, and United Arab Emirates.

Bräutigam, D. 2009. *The Dragon's gift. The real story of China in Africa*. Oxford: Oxford University Press. A readily accessible book on China-Africa relations by the most outstanding scholar in the field. It demystifies China's engagement on the continent and provides an informed response to the politicised debate on Sino-Africa relations.

Carmody, P. 2011. *The new scramble for Africa*. Cambridge: Polity Press. An important discusson of Africa's new role in global geopolitics. It offers insights into how and the extent to which resources matter for Africa's development and what role 'emerging' economies play in this.

Fejerskov, A. M., E. Lundsgaarde, and S. Cold-Ravnkilde. 2017. Recasting the 'New Actors in Development' Research Agenda. *The European Journal of Development Research* 29 (5):1070–1085. This review article questions the analytical value of distinguishing between 'new' and 'old' development actors as heterogenity between actors seems to cut across these boundaries and calls for explanations of how SSC differ from North-South Cooperation beyond what can be explained by scrutinising sector focus, scope, and vestors of engagement.

Mawdsley, E. 2012. *From Recipients to Donors. Emerging Powers and the Changing Development Landscape*. London: Zed Books. A clearly written monograph on the 're-emergence' of a number of new development actors. It lays bare these re-emerging donors' history and offers very valuable insights into the effects of this trend for development cooperation.

Richey, L. A., and S. Ponte (eds.) 2015. *New actors and alliances in development*. New York: Routledge. An edited volume that brings together a range of scholars who analyse how development financing is changing and how and to what extent it matters for development 'as we know it'.

2 South-South Cooperation in historical perspective

The end of colonialism, the Bandung Conference, and the Non-Aligned Movement

The end of World War II signalled a new world order. On the one hand, several institutions were established to govern the world economy and make sure that the world would not experience another war on that scale. These institutions included the International Bank for Reconstruction and Development (now the World Bank), the International Monetary Foundation, and the United Nations. On the other hand, the end of World War II kick-started formal decolonisation in Asia and Africa (most of Latin America was decolonised in the first quarter of the nineteenth century). It began in Asia just after the war (Indonesia and Vietnam in 1945, Burma and Ceylon in 1948, and Cambodia in 1953 for instance) and continued in Africa (e.g. Tunisia and Morocco in 1956, Ghana in 1957, Cameroon, Togo, Mali, and Niger in 1960, Sierra Leone in 1961, and Rwanda and Uganda in 1962). However, the new institutions established to regulate world affairs did not take into consideration the reality of the numerous independent countries that saw the light of day in the decades that followed World War II. Therefore, the early Cold War years were also characterised by a quest from developing countries for a greater voice in international affairs.

The first major episode in this quest was the Conference on Afro-Asian Peoples held in the city of Bandung, Indonesia in 1955, normally referred to as the Bandung Conference. The conference, convened by Indian Prime Minister Jawaharlal Nehru and Indonesian President Sukarno, brought together representatives from 29 nations and colonies in the decolonising Global South. The idea was to give voice to this part of the world, but also to present a non-aligned bloc amidst the growing East-West conflict in the Cold War.

The 29 countries represented approximately half of the world population and they wanted to develop policies to exert influence on world politics and resist imposition from others. The final Communique of the Bandung Conference called for 'self-determination' and for world politics to be based on equality of races and nations. Also of importance were human rights and challenges faced by the so-called dependent countries, i.e. countries that were still ruled by their colonial powers. The Bandung Conference did not appear out of nowhere. Rather, it grew out of the national movements in Asia and Africa fighting European imperialist and racial policies. Hence, it comes as no surprise that the final communique declared that the participants were against imperialism in all its manifestations.

This fight against imperialism and neo-colonialism became part and parcel of the Non-Aligned Movement (NAM) that grew out of the Bandung Conference. NAM was established in Belgrade in 1961. It aimed to give developing countries a collective voice and preserve their political independence in a world increasingly affected by the nuclear race of the 1950s and 1960s. The idea of NAM as a middle course between the two superpowers was originally conceived by the then heads of states of Egypt, Ghana, India, Indonesia, and Yugoslavia and put forward in the UN but it was not until the first summit in 1961 that an alternative forum was created for negotiating diplomacy and autonomy in the new world order. NAM grew out of the de-colonisation process and was launched as an alternative to the Western-dominated world order guided by the UN. While NAM originally was concerned with political effects of the Cold War, the focus, aim, and membership of NAM soon broadened to include anti-colonialism, challenges of economic development, and inequality.

Each NAM meeting resulted in a resolution highlighting the main conclusions from the meeting. The Lusaka Declaration (1970) turned out to be of great significance for SSC. It centred on the concept of 'self-reliance', i.e. economic cooperation among developing countries, which was further developed at the Georgetown Conference in 1972 and fine-tuned at the NAM Algiers Conference in 1973 where specific areas of cooperation were singled out to further self-reliance. These areas included trade, industry, and technology and training. NAM was also important in relation to SSC as it was the foundation stone of the call for a New International Economic Order (NIEO) (see p. 36) and as it brought SSC into the UN system

(Morphet 2004; Tomlinson 2003; Folke, Fold, and Enevoldsen 1993; Bandyopadhyaya 1977; Cox 1979).

In 1973 NAM's call for SSC and increased trade and investments led to the establishment of The Working Group on Technical Cooperation among Developing Countries in the United Nations Development Program (UNDP). The main aim of the working group was to share experiences and transfer knowledge among developing countries to enhance the effects of development aid, and further regional technical cooperation among developing countries. The group was made up of members from both the Global North and the Global South. It was closely related to the call for a NIEO and challenged the donor-recipient division already existing in bi- and multilateral engagements between developed and developing countries. The group argued that the necessary competencies and capabilities already existed in the Global South – what was needed was to make these resources available. It also came up with a number of recommendations including a plea to developing countries to develop their own technical cooperation (i.e. aid) programmes; a request to the UNDP to make use of local capacities in developing countries and to channel money via regional entities; a call to establish a special unit within UNDP to coordinate South-South activities; and finally – in line with the political heritage of the NAM – a call for support for national liberation movements in the Global South.

The working group led to the establishment of a special unit in the UN system still in existence, the UN Office for South-South Collaboration, that mainstreams SSC in the UN system; engages members and external stakeholders to support SSC; pushes new SSC ideas to members; and manages the 'UN Fund for South-South Cooperation' as well as the 'Pérez-Guerrero Trust Fund for Economic and Technical Cooperation among Developing Countries' (UNDP 1974).

The UN Office for South-South Collaboration is also closely related to the 1978 Buenos Aires Plan of Action for Promoting and Implementing Technical Co-operation among Developing Countries (UNDP 1994). The Buenos Aires Plan of Action is the outcome of the 1978 UN conference held in Argentina that aimed to enhance global partnership for development. The conference was held between the two oil crises that divided the Global South and led to renewed focus on national and collective self-reliance (see p. 39). The Buenos Aires Plan of Action was important as it gave first priority to SSC and perceived technical cooperation among developing countries as the means to reach self-reliance leading to the creation, acquisition, adaption, transfer, and pooling of knowledge and experience that mutually benefits the countries of the Global South and which ultimately leads to social and economic development. Importantly, the plan adopted a very pragmatic view on SSC: technical cooperation among developing countries was perceived to take a multitude of forms and include a multitude of actors, from multilateral to regional and national organisations and both public and private entities. Notably, it was suggested that it focuses on methods and techniques that were adapted to local needs.

South-South economic integration

Despite the political development towards closer SSC, South-South economic integration was still minuscule in the 1960s and 1970s. South-South trade, for instance, only made up some 4–7 per cent of world trade from 1950–1975 (Folke, Fold, and Enevoldsen 1993; UNCTAD 2015) and although both manufactures and primary commodities were traded across borders in the Global South, the majority of the products traded within the Global South were fuels, food, and other raw materials (see figure 2.1). The two oil crises of the

Figure 2.1   South-South trade in select product groups as percentage of world exports

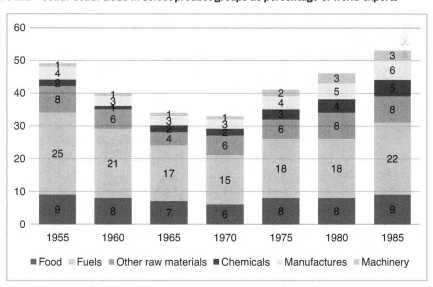

Source: (Folke, Fold, and Enevoldsen 1993: 13)

1970s (see p. 39) effectively stopped South-South trade. Hence, in the 1980s South-South trade only made up some 3–5 per cent of world trade. Only by the late 1990s did South-South trade begin making inroads into total world trade – growing from 5 per cent in 1998 to 15 per cent in 2013 (Kristinsson 2017: 18). China, of course, accounts for a large share of this increase.

Likewise, South-South investments were microscopic compared to world FDI in the 1960s and 1970s. This is somewhat surprising as transnational corporations from the Global South developed in the 1960s. However, these new 'third world multinationals' in no way covered all of the Global South – neither in terms of origin nor in terms of markets targeted. Rather, they predominantly originated in Latin America (and to a lesser extent India) and focused on investments in neighbouring countries. Moreover, they did not change the overall division of labour between the North and the South as they primarily targeted the primary sector; and their competitive advantage was to a large degree associated with policies of import substitution and regulation of investments (Gammeltoft 2008; Ramamurti 2009). In short then, the political developments of SSC in the decades that followed WW2 were not mirrored by economic developments. In contrast, South-South economic cooperation seemed to stagnate:

Bilateral South-South Cooperation

Aggregate economic data, however, tend to hide the growing importance of bilateral collaboration between countries in the Global South, such as Cuba's engagement in Angola (see Textbox 2.3); China's engagement in most of the countries on the African continent (see Textbox 2.1); and India's engagement first in neighbouring countries and then in Africa. Bilateral collaborations were to a large extent driven by newly established aid programmes in parts of the Global South. Important in this regard was China and India (and their mutual fight for recognition), but it also included the Gulf donors and a number of countries belonging to the former Eastern Bloc (Mawdsley 2012b; Kragelund 2008).

China's aid programme grew out of the Bandung Conference and in the 1950s China expanded its foreign policy concerns from Asia (in particular the Korean peninsula) to also include the African

Box 2.1 The TAZARA railway

On 12 July 1970 the Presidents of China, Tanzania and Zambia signed an agreement to build an 1,860 – kilometre railway from Kapiri Mposhi in Zambia to Dar es Salaam at the coast of Tanzania. The construction, materials, trains, and railway carriages were all funded by a Yuan 988 million (USD 415 million) interest-free loan from China and some 30–50,000 Chinese railway experts (alongside approximately 80,000 Tanzanian and Zambian workers) were in charge of the construction.

The idea of such a railway was not new. Already prior to independence, Julius Nyerere and Kenneth Kaunda (to become first presidents of Tanzania and Zambia, respectively) discussed the possibility of connecting the Zambian Copperbelt to the Indian Ocean and thereby on the one hand, relieve Zambia from its reliance on the white dominated South Rhodesia, South Africa and (Portugese) Mozambique (during this time all copper was exported either via the port of Beira, Mozambique, or the port of Durban, South Africa) and on the other hand, develop the southern regions of Tanzania by connecting it to the coast.

However, neither Nyerere nor Kaunda had been able to persuade foreign donors that the railway was worth supporting. For instance, both the United Nations and the World Bank concluded that the railway was not economically sustainable and hence were not willing to fund the project. Instead, Tanzania and Zambia approached China – first, during Nyerere's visit to China in February 1965 and again during Zhou Enlai's, China's Premier, visit to Tanzania the same year in June. During these two visits, it was agreed to make another appraisal of the railway – this time led by a Chinese team of experts. The result of this appraisal led to the formal agreement in September 1967 to pursue the construction of what was to become China's largest foreign aid project and the third largest infrastructure project in Africa.

Unlike a potential railway funded by the World Bank and constructed by for instance European firms, the railway funded and constructed by China made extensive use of manual labour and capital equipment manufactured in China. The reason was simple: China was a developing country like her two African counterparts and while labour was not a problem, cash – and in particular, hard currency – was hard to come across. Therefore, China also invented a so-called commodity credit agreement that was to finance local costs in Tanzania and Zambia (52 per cent of total costs). Essentially, this agreement entailed that Chinese textiles, pharmaceuticals, toothpaste, housewares etc. worth the local costs were to be sold in government cooperative shops in the two African countries and profits from the sales were then used to pay for local wages, housing, food etc. While this maybe sounds of little importance, it was a major contributing factor to the quadrupling of imports from China to Tanzania (increasing from 5 per cent of total imports in 1969 to over 22 per cent in 1973).

While the project was scheduled to end in 1977, the construction workers managed to complete TAZARA in 1975 – two years ahead of time – and passenger services on the train began a year after, in 1976. This was perceived as an immense success for Chinese aid: not only did they manage to build the railway in very tough terrain – thereby simultaneously fighting white supremacy in Southern Africa and supporting pan-Africanism – they also did it ahead of schedule. However, running the TAZARA railway turned out to be less successful: the Chinese-built locomotives often broke down; they lacked the power to haul copper through the mountainous regions; heavy rains led to landslides that destroyed parts of the physical infrastructure; offloading and shipment was slow in Zambia; and the port of Dar es Salaam was inefficient. The net effect was large-scale operating losses for the railway. In order to change this, first China and then several 'Western' donors chipped in to replace the Chinese locomotives with German ones, to buy spare parts, and pay for technical experts to assist in repairing and maintaining the equipment.

Moreover, the interest-free loan fixed in Yuan with an initial grace period of five years (later extended) and repaid in 'hard' currencies over a 30 – year period turned out to be less cheap than first anticipated due to exchange rate fluctuations: the Tanzanian shilling, for instance, devalued rapidly vis-à-vis the US dollar at the beginning of the 1970s while the Chinese Yuan appreciated. This essentially meant that the local value of what was to be repaid almost doubled.

Notwithstanding these challenges, TAZARA is still perceived as a poster-child of South-South Cooperation. It was the largest single development finance project financed by a developing country (and is still by far the biggest single project financed by China); it was agreed at the height of the Cold War and allowed Tanzania and Zambia to reject aid from Washington as well as from Moscow; it put the eight principles governing Chinese aid into practice for instance by providing interest-free loans, by helping the recipients to embark on a development road based on self-reliance, by yielding quick results, by training local personnel, and by making sure that the Chinese experts enjoyed the same standard of living as the local experts. For the US and Western Europe, in contrast, it was perceived as the 'great steel arm of China'.

Sources: (Monson 2010; Mwase 1983; Bailey 1975)

---

continent. The cornerstone of China's external relations were the five principles of peaceful co-existence formulated jointly by China and India in 1954 and signed a year later by the 29 participants during the Bandung Conference. In short, China argued that China and the African countries belonged to the same club of developing countries. These principles were the foundation for '*The Chinese Government's Eight Principles for Economic Aid and Technical Assistance to Other Countries*' (henceforth: the Eight Principles of Chinese Assistance) that were negotiated during Premier Zhou Enlai's trips to African and Asian countries in 1963–1964 and finally signed in January 1964 in Ghana. The Eight Principles of Chinese Assistance are:

1   Aid is provided on the principles of equality and mutual benefit.
2   China respects the sovereignty of recipient countries and never attaches conditions to its aid.
3   Aid is provided as interest-free or low-interest loans with extended time for repayment (if necessary).

4   The overall aim of Chinese aid is to make recipient countries self-reliant step-by-step.
5   China's projects are characterised by low costs and fast delivery times.
6   Chinese projects use equipment and material of the best quality at international market prices manufactured in China. The Chinese government undertakes to replace equipment if it is not up to standards.
7   Chinese aid seeks to build capacity in recipient countries to fully master the techniques related to the aid project.
8   Chinese experts will enjoy the same standard of living as the experts of the recipient country.

Although these principles have guided Chinese aid relations since, they do not provide the full picture. Chinese development finance is not only guided by mutual benefit and equality, it is also driven by geopolitics (e.g. fighting white supremacy in Southern Africa (see also Textbox 2.1), getting a permanent seat in the UN Security Council, fighting Moscow and Washington during the Cold War (see also Textbox 2.3), and supporting allies in the Middle East); economics (e.g. paving the way for internationalisation of Chinese firms (Bräutigam and Xiaoyang 2011)); getting access to resources (Corkin 2013; Jansson 2013); and as we have seen in chapter 1, a quest for international political power that mirrors its growing economic power. India's aid programme to a large extent mirrors China's. It also began in the 1950s – first targeting neighbouring countries and then also targeting specific countries outside the region. It grew out of the same solidarity concern as the Chinese programme but soon came to be influenced also by competition with China, access to new markets, energy security, and military concerns (Cheru and Obi 2011).

The two Asian giants, however, were not the only Southern countries that began to use development finance to establish closer relations to other countries of the Global South. This was also the case for a number of OPEC countries including Kuwait that established the Kuwait Fund for Arabic Economic Development in 1961. United Arab Emirates and Saudi Arabia established similar funds in 1971 and 1974, respectively (see also Textbox 4.2). Aid from these OPEC countries was intended to spearhead a new international order based on regional self-sufficiency (Villanger 2007; Kragelund 2008).

G77, UNCTAD and the New International Economic Order

What we saw then in the post-World War II years was that bi- and multilateral South-South Collaboration developed in tandem. Ideas fostered at grand meetings fed into how bilateral engagements were governed and growing bilateral engagement called for increasing international governance of SSC. Therefore, it came as no surprise that in parallel with NAM and growing bilateral relations, 77 developing countries in 1964 formed a pressure group within the UN system '*to articulate and promote their collective interests and enhance their negotiating capacity . . . within the UN system, and promote South-South cooperation for development*' (G77 2017). The Group of 77 (G77), which got its name from the number of founding countries, now comprises some 130 membership states (some of which are now members of the EU and therefore do not (solely) represent the South) and is by far the largest grouping of Southern countries. The formation of G77 was closely linked to the first UN Conference on Trade and Development (UNCTAD) and soon the G77 became the prime mover in pushing South-South Cooperation issues within the UN system as well as the main vehicle to further the economic and political rights of the Global South.

Through production of new information and the inclusion of heterodox economics into the development arena (see Textbox 2.2) UNCTAD came to influence development thinking and doing throughout the Global South. Most importantly, UNCTAD brought the issue of non-reciprocal tariff preferences for manufacturing exports of developing countries by developed countries to the fore. The aim of the tariff preferences was to push import-substituting industries in the Global South to export to the Global North. Although the effects of the tariff preferences were significantly watered down due to the exclusion of agricultural products and the fact that labour-intensive manufactured products were to be negotiated on a case-by-case basis, this agreement signed in 1968 is perceived as one of the greatest achievements of UNCTAD (Karshenas 2016).

Box 2.2 The emergence of development economics

What is now known as development economics originated in the empirical problems encountered by economists in Latin America in the 1930s (in particular Raul Prebisch and later Carlos Furtado), and in Western Europe in the aftermath of World War II. Here economists struggled with the apparent mismatch between the neoclassical theories and the reality on the ground. For Latin American economists, the main concern was the collapse of international trade that led to foreign exchange and import controls and for Western European (and to some extent American) economists focus increasingly was on 'underdeveloped' regions of the world. For Latin America as well as for Asia and Africa (after decolonisation) focus turned to policy issues on how to bring about economic development in these regions.

In addition, the establishment of the UN system and the combination of availability of new data from the Global South and inspiration from a range of different academic traditions were important factors in furthering the development economics sub-discipline. The development of specialised bodies within the UN system meant that focus moved away from saving rates and economic development to also include other aspects of development in the Global South. Moreover, it came to include aspects not dealt with before such as the informal sector and inequality. Concomitant to this, the UN system itself generated new data and information on development that could be used to analyse new aspects of development. In particular, survey data on employment, consumption, and savings were made available to academics and policy makers alike. Finally, academics like Nicholas Caldor, Simon Kuznets, Arthur Lewis, Gunnar Myrdal, and Hans Singer brought ideas from other disciplines into economics.

The result was that focus increasingly turned away from orthodox international trade theory to analysing underlying structural characteristics of developing countries and how these structures affected development; issues of unequal power relations; deteriorating terms of trade for commodity exporters; the need for protection of domestic industries; and the international trade regime. This led to a focus on issues such as technological change, economies of scale, appropriate technologies, structural transformation, commodity agreements, and tariff preferences.

The experiences of Argentina (Prebisch) and Brazil (Furtado) led to a rejection of traditional trade theories (e.g. comparative advantage). Instead, the Latin American structuralists advocated for import substitution industrialisation strategies that could lead the way for economic transformation in the Global South. Structuralist economic thinking was not confined to Latin America and the Economic Commission for Latin America. The rejection of neoclassical economic assumptions and the call for national economic policy formulation was soon adopted outside the continent, for instance by development economists like Dudley Seers, Hans Singer, and Gunnar Myrdal. Development economics of this era established that there is a difference between economic growth and economic development; that economic development is related to technological advances; that structural transformation matters for development; and that underdevelopment is closely linked to the role of the Global South as commodity exporters in the world economy.

Sources: (Hunt 1989; Thorbecke 2007; Karshenas 2016).

Equally important, the General Assembly of the United Nations in 1974 ratified the Charter of Economic Rights and Duties of States and the NIEO. The declaration of a New International Economic Order was a proposition for a new economic and political framework for international relations between equal countries. In many ways, the declaration of a NIEO was the culmination of two decades of struggle for a new world order that would give more power to developing countries to regulate and control economic activities in their own territories. The name itself had three major connotations: first, it referred to the problems of the existing economic order of the day; second, it blamed Europe and the rest of the Global North for the situation; and consequently, third, it pointed to a major restructuring of power in order to overcome the problems[1].

In short, the NIEO was a call for a break with the existing international division of labour where industrial production was concentrated in the Global North that imported raw materials and cheap labour from the Global South and then exported the manufactured products back to the Global South. In other words, the development in the affluent part of the world was linked to the 'underdevelopment' of the rest of the world and seen from the perspective of the Global South, the Global South was losing out.

The then Secretary-General of UNCTAD, Gemani Corea, explained the need for a NIEO to an audience at the London School of Economics in 1976 in the following way:

> *The two and a half decades that spread over the conclusion of the Second World War and the beginning of the 1970s witnessed vast acceleration in the economic and productive capacity of the developed countries . . . It witnessed a remarkable upsurge in the intra-trade of these countries. But the balance sheet for the developing countries was not similarly impressive. Their economic expansion did not suffice to meet the needs of the growing populations, to overcome in a significant way the problems of mass poverty, malnutrition, and unemployment. Their share in world trade declined during this period . . . In the period since the beginning of the 1970s the prevailing order has ceased to work well . . .*
>
> (Corea 1977: 178f).

In particular, the proponents of NIEO pointed to the deteriorating terms of trade, trade imbalances, lack of control of transnational corporations' activities, and lack of technology transfer as core issues justifying the NIEO[2]. In short, the NIEO revolved around a demand for structural change in global power relations and called for self-reliance among developing countries. The structural change included a demand for a reform of the Bretton Woods institutions; equality among states; access to industrial countries' markets for manufactures from the Global South; establishment (and enforcement) of institutions to correct global balance of trade imbalances and debt problems; and power to developing economies to regulate and control transnational corporations' activities. Self-reliance included giving full sovereignty to states over natural resources and thus the right of nationalisation of transnational corporations; promotion of endogenous technological development; establishment of primary commodity associations; and strengthening of SSC (Corea 1977; Golub 2013; Fröbel, Heinrichs, and Kreye 1978; Cox 1979).

The temporary collapse of South-South Collaboration: the Cold War, the oil crisis, and the debt crisis

In the years that followed the launch of the NIEO at the UN General Assembly in April 1974 the countries of the Global South engaged in negotiations with representatives of the Global North over these subsets of the NIEO, but even though the proposition received much attention among politicians, practitioners, and academics, it never came to make any real difference on the ground. The NIEO therefore in many ways was the culmination of the first SSC era.

Several related incidents meant that the NIEO lost momentum before it really took off. They included the intensification of the Cold War that led to increasing alignment to a superpower – rather than to a Southern partner; the break-up of the South into the affluent oil-rich countries and resource-poor developing countries; decreased willingness to finance the redistribution implicit in NIEO due to the economic crisis and the ideological shift towards neoliberalism in the Global North; and recycling of petrodollars, i.e. the international spending or investments of an oil producing country's revenues from the export of petroleum, that eventually led to the debt crisis. The boom in oil prices massively increased the availability of petro-dollars that were lent to non-oil producing countries to finance the rapidly increasing costs of importing oil.

Box 2.3 The Cubans in Angola

Over one and a half decades beginning in the mid-1970s and ending in 1991, Cuba dispatched tens of thousands of soldiers in Angola. At the height of the Angolan civil war in 1988, Cuba had dispatched more than 50,000 soldiers in the country. No other developing country projected its military power across continents as Cuba did during the height of the Cold War. According to Gleijeses (2006: 44) Cuba also dispatched more than 70,000 aid workers to Africa in this period.

Cuba's Africa journey, however, did not begin with the civil war in Angola. In fact, Cuba was present in Africa from just after the Cuban Revolution in 1959 and Cuba was a critical factor in the independence war in Guinea-Bissau from 1966 to 1974. Likewise, it did not end in Angola. Roca (1980: 59) estimates that in 1978 roughly 4,500 Cuban construction workers were busy erecting bridges, building airports and schools mostly on the African continent. On top of this, Cuban medical doctors and dentists were busy setting up (and running) medical centres across the Global South and Cuban teachers (at primary school level as well as at the level of higher education) were sent to African countries to teach and set up education systems. Furthermore, Cuba engaged in trade union organisation, agricultural sector developments (irrigation, fishing, sugar, coffee etc.) and development of the transport sector. Roca (1980: 57f) reckons that Cuba in 1978 had posted just under 40,000 military advisers and troops across Africa (mostly Angola (19,000) and Ethiopia (16,500)) and over 10,000 civilian personnel to Africa. A decade later, in 1986, the total number of Cubans that had served in Africa had reached 250,000 (Benzi and Zapata 2017: 87).

Angola (and the rest of Africa) was of interest to Cuba for a number of reasons. In line with the Bandung Conference and NAM, Cuba supported the liberation movement across Southern Africa. Thus, the official aim of Cuba from the 1970s onwards was to assist national liberation movements in fighting white supremacy and colonialism. Economics and geopolitics, however, were also of importance. In terms of economics, Southern and Central Africa were rich in oil and minerals and could potentially provide Cuba with much needed resources. Linked hereto, Cuba's foreign assistance programme from the mid-1970s onwards was based on the 'ability to pay' principle entailing that poor countries only had to pay food and shelter for the Cuban personnel (and the Soviet Union funded infrastructure and equipment) while more affluent countries were charged for the services. This meant access to much needed convertible foreign exchange. Benzi and Zapata (2017: 87), for instance, estimate that Cuba's international engagement generated approximately USD 50 million in 1977 alone. More importantly, however, Cuba's intervention in Angola and Ethiopia meant increased leverage with the Soviet Union as it enabled Cuba to reciprocate Soviet economic commitment in Cuba. Geopolitics also mattered in another way. Countries like Angola and Mozambique were of geostrategic importance due to their long coastline to the Atlantic Sea and Pacific Ocean and their access to central Africa via the Congo River as well as their status as frontline states vis-à-vis the white dominated Southern Africa. Most importantly, probably, Castro had to fight back against the US. It would have been suicidal to fight the US directly. Hence, Castro had to fight its northern neighbours indirectly: supporting the Movement for the Liberation of Angola (MPLA) in Angola did exactly that by simultaneously gaining friends and weakening US influence in the region. Finally, history mattered: the majority of Cuba's population is of African descent.

Plate 2.1 *Former Cuban President, Fidel Castro, commemorating soldiers that died in Angola, Havana, July 1989*

In other words, while the official aim of Cuba's military involvement in Angola was the latter's national liberation, economic and geopolitical reasons at home were equally important. Importantly, the involvement also resulted in a South-South conflict. Havana sided with Moscow in supporting the MPLA while Washington supported the National Liberation Front of Angola (FNLA) and Beijing first supported FNLA and then the National Union for Total Independence of Angola (UNITA). In effect then, China and Cuba (alongside the Soviet Union) were supporting opposing sides in the conflict. More decisive for SSC, however, it came to the knowledge of the African peoples that by the mid-1970s South Africa had joined the conflict siding with China. At the end of the day, then, Cold War politics and Beijing's conflict with Moscow was considered more important than the official rhetoric of solidarity with the African peoples.

Cuba's massive presence in Angola ended in December 1988 when it was agreed that South Africa would end its support to UNITA and that the Cuban troops would leave Angola approximately two years later. In 1991, the last Cuban troops left the Southern African country.

Sources: (Falk 1987; Gleijeses 2006; Roca 1980; Gleijeses 1997; Benzi and Zapata 2017).

As we have seen, SSC grew fast from its outset in the mid-1950s. SSC soon became formalised and ever more organisations adopted it. In parallel, however, the Global South experienced a growing internal economic and political differentiation – especially in the 1970s cf. (Menzel 1983). Economically, industrialisation processes took off in some countries in the Global South while others continued to rely on exports of raw materials (see also chapter 3); and considerable price fluctuations of primary commodities affected commodity exporters and commodity importers differently. Oil-exporting countries obviously gained from the increase in prices while oil-importing developing countries lost out due to the higher prices as well as due to the recycling of petrodollars that led to the easy availability of cheap private loans at floating interest rates. This balance of payment crisis in oil-importing economies led to growing debt that later became the cornerstone of the 1980s' debt crisis, which further undermined the idea of a collective South. Moreover, the debt crisis initiated a process of de-industrialisation that created serious setbacks, especially in Africa and Latin America, which again negatively influenced South-South economic cooperation.

Politically, the period also witnessed the growth of regional power blocks and mounting ideological differences. Likewise, the relatively broad (trade) agenda of the GATT Uruguay Round (1986–1993) made political disagreements of the developing countries come to fore, thereby effectively leading to a division among developing countries and, thus, a crisis of the G77 (Folke et al., 1993; Lima & Hirst, 2006).

The result was that the common history of colonialism, common social and economic structure, common interests, and the common 'enemy' (i.e. the Global North) were not sufficient to keep the Global South together. Rather what we witnessed was an increasing process of differentiation that led to the end of the first epoch of SSC.

Conclusion

This chapter has shown that SSC is in no way a new phenomenon. Rather, most of the current SSC engagements refer back to the 'original' SSC. The actors and institutions use the same concepts and frame their engagement along the same arguments. This chapter has also shown that despite a common vocabulary and a common 'enemy' the Global South did not manage to stick together in times of rapidly changing geopolitics and economics. Instead, the great heterogeneity came to the fore leading to an 'emerging South' and 'the rest of the Global South'. Thereby, the political platform their demands rested upon disappeared.

Discussion questions

- Consider the main reasons for the origin of South-South Cooperation.
- Discuss why South-South Cooperation thrived politically despite the lack of economic integration in the Global South.
- Explain the main motives for a New International Economic Order and why it collapsed before it really took off.

Web pages of interest

- The Group of 77 is the official website of the UN's group of 77. It contains the official documents and news related to the group: www.g77.org/
- United Nations Office for South-South Cooperation is the official UN website for SSC. It contains information on the UN's work on SSC and triangular cooperation. It includes lots of resources and news: www.unsouthsouth.org/
- United Nations Conference on Trade and Development is the UN's main body dealing with aspects of economic globalisation. It offers good data on trade and investment: unctad.org/en/Pages/Home.aspx

Notes

1 For many commentators the NIEO was not as new as the name indicated. Rist (2008: 151), for instance, puts it this way: '*Despite its* [NIEO] *rhetorical list of demands, it proposes nothing new – or anyway, nothing that sheds fresh light on how to improve living conditions for the peoples of the South.*' He argues that the Southern proponents of the NIEO just wanted to ensure that the elite of the Global South received a larger share of the economic growth via more aid and access to their own resources.

2 As Cox (1979) reminds us, the proponents of NIEO in the first half of the 1970s were many and included academics in the Global North writing from orthodox perspectives but who still called for 'correction of inequalities among countries'; social democratic inspired commentators that also stressed the needs of the poor; a diverse group representing the Third World (including Corea of UNCTAD) focusing in particular on intellectual and

material self-reliance; a 'neo-mercantalist' group that paid particular attention to how economic power led to political power; and finally, a Marxist, historical materialist inspired group focusing on the intrinsic link between development and underdevelopment – between core and periphery.

Further reading

Golub, P. S. 2013. From the New International Economic Order to the G20: how the 'Global South' is restructuring world capitalism from within. *Third World Quarterly* 34 (6):1000–1015. This article analyses how the Global South sought to build an institutional framework in the 1960s and 70s, why it failed, and to what extent history is now repeating itself.

Monson, J. 2010. *Africa's Freedom Railway: How a Chinese Development Project Changed the Lives and Livelihoods in Tanzania.* Indiana University Press: Bloomington. A very rich description of the encounter between Chinese railway workers and their Tanzanian counterparts in the construction of the TAZARA railway from Dar es Salaam in Tanzania and Kapiri Mphosi in Zambia. It shows how Chinese aid principles were put into practice in the 1970s.

Rist, G. 2008. *The history of development: from western origins to global faith (3rd edition).* London: Zed Books. An exceptional overview of the history of development and how the concept has changed in the past 70 years.

3

The resurgence of South-South Cooperation

Sustained economic growth in 'emerging' economies

At the time of the Bandung Conference (see chapter 2) developing countries in South East and East Asia were performing economically on a par with African countries (to become independent in the years that followed), and far worse than their counterparts in Europe and the Unites States. However, only a decade later – by the mid-1960s – the picture changed: not only did growth rates in South Korea, Taiwan, Hong Kong, and Singapore, the first-tier Newly Industrialised Countries (NICs), rise rapidly, they also continued their upward movement beyond an initial inflow of external finance (approx. 5–6 %/year from 1950–1975). Moreover, growth rates in the NICs outperformed growth rates in Europe and the US. Alongside this rapid and sustained growth, the economies of the NICs also transformed structurally. From being based predominantly on extraction of primary commodities, they developed high performing manufacturing and service sectors. Alongside these changes, standards of living began to increase rapidly, the level of education improved dramatically, and companies from these countries began to internationalise (Chang 2003; Lauridsen 2008).

Taken together these changes meant that a challenge to the world division of labour 'as it was' where countries of the Global South exported primary commodities to the Global North that manufactured these commodities and re-exported (some of) them back to the Global South, thereby benefitting from the greater value addition in the manufacturing process. As Table 3.1 shows, manufacturing production grew at a far greater pace in the first-tier NICs compared to both European countries and countries in Latin America. This meant that by the mid-1990s, South Korea had overtaken a manufacturing powerhouse like Brazil in terms of share of world manufacturing production and Taiwan was six times as important worldwide as Greece in terms of manufacturing.

Table 3.1 *Growth of manufactured production, 1963–1994*

	Share of world manufacturing output (%)			Average annual change (%)		
	1963	*1980*	*1994*	*1960/70*	*1970/81*	*1980/87*
Hong Kong	0.1	0.2	0.2	ND	10.1	ND
Singapore	0.1	0.1	0.3	13	9.7	3.3
South Korea	0.1	0.6	2.7	17.6	15.6	10.6
Taiwan	0.1	0.5	1.2	16.3	13.5	7.5
Argentina	ND	0.8	1.5	5.6	0.7	0
Brazil	1.6	2.4	2.6	ND	8.7	1.2
Greece	0.2	0.2	0.2	10.2	5.5	0
Spain	0.9	1.7	1.4	ND	6	0.4

Source: Adapted from: Dicken (1998: Table 2.3)

Not only did these changes challenge the established view of the international division of labour, they were furthermore an attestation that the Global South was indeed not a homogenous unit. As we saw in chapter 2, the two oil crises of the 1970s divided the Global South into an increasingly affluent oil-exporting and an increasingly indebted and poor oil-importing group of countries. South East Asian countries' economic transformation from the 1960s onwards added a further dimension to this heterogeneity. The Global South now comprised economies primarily based on primary exports as well as economies increasingly based on manufacturing and services.

Box 3.1 The Newly Industrialised Countries

A number of countries in East and South East Asia experienced rapid economic growth in the post-World War II years. This growth was driven by rising exports of manufacturing goods. Due to their rapid economic growth and geographical origin, these countries were often referred to as *Asian tigers* or *tiger economies*. Neither of these terms, however, informed us about the causes of the economic growth. In contrast, the term NICs directs our attention to the underlying causes of this growth, namely the structural transformations that took place in these countries.

A number of other countries also experienced exceptionally high growth rates of manufactured goods in the 1970s. They included among others Ivory Coast, Jordan,

Mauritius, Sri Lanka, and Tunisia. In contrast to these countries, the growth rates of the NICs continued beyond the 1970s, and the rising exports of manufacturing goods were associated with a number of other changes, including internal division of labour, the rise of so-called 'third world multinationals', high savings and investment ratios, rising standards of living, and rural-urban migration.

A key difference between the NICs and other exporters of manufacturing goods from the Global South was that the NICs quickly managed to move beyond (only) labour-intensive manufacturing (textile yarns and fabrics, clothing, and light manufacturing) to also manufacture and export transport equipment, electrical machinery, and iron and steel products. This was only possible due to higher levels of education, higher wage levels to create domestic demand, and better infrastructure. Normally, we distinguish between the first-tier NICs (Hong Kong, Singapore, South Korea, and Taiwan) and second-tier-NICs (Thailand, Malaysia, Indonesia, and the Philippines).

The NICs sparked two important discussions. First, they initiated a long-term discussion of how these countries managed to transform their economies and pave the way for sustained growth (barring the disruption caused by the 1997 Asian financial crisis). While the international financial institutions led by the World Bank argued that the developments we have witnessed in the NICs were primarily driven by market-led economic policies of openness, heterodox economists pointed to the central role of the (developmental) state in bringing about this development. This latter group argued that the structural transformation that we witnessed in the NICs was not market-driven. In contrast, it was driven by strong political leadership; collaborative state-business relations; industrial policies etc.

Second, they revived the discussion of categorisation of the world in categories such as: first, second, and third word, and developed and developing countries (see Textbox 1.1). The NICs demonstrated that economies were not necessarily stuck in underdevelopment; it was indeed possible to embark upon a road of sustained economic growth. Simultaneously, their developments revealed how long the journey from developing to developed country status is and that this binary idea of the world misses out the heterogeneity within both groups.

Sources: (Lauridsen 2008; Leftwich 1995; Chang 2003; Tan 1993)

Emerging growth of political influence

Although the NICs influenced understandings of income distribution in the world, they did not really rock the boat in terms of who is in power – and who is not. Both first- and second-tier NICs were simply too small politically and economically to change our perception of world politics. No doubt, NICs shaped particular sectors in the world. The growth of *chaebols* like Samsung and Hyundai in South Korea, for instance, transformed the market for LCD screens and autos, respectively. Likewise, Taiwanese tech companies transformed the market for integrated circuit market as well as the market for mobile phones for many years (Farooki and Kaplinsky 2012). NICs' economic transformation was thus linked to their increasing participation in the global economy but they did not change the terms of interaction.

This was also the case for China. China's economic growth in the 1980s was largely based on imports of capital equipment and raw materials that were transformed into light consumer goods and re-exported to the world market. Gradually, Chinese companies took in new technologies and began producing and exporting more complicated goods, including capital equipment. This transformation not only meant that production of manufactured goods moved from one part of the world to another, the economic activity also sparked the growth of Chinese consumers that led to a rapidly growing domestic Chinese demand that further fuelled the economic growth process in China. The scope and speed of this process meant that by the turn of the millennium, the centre of gravity of consumption was shifting from Europe and the US to China (Farooki and Kaplinsky 2012).

The big difference between the impact of the NICs on the world economy and China's impact was the size of the Chinese economy. While the population of the NICs made up less than five per cent of world population, China's population makes up approximately one-fifth. Thus, China's demand for primary commodities to fuel 'the factory of the world' affects commodity prices and thus economic development all over the Global South. Likewise, the increasing Chinese demand for, among other things, consumer goods alters price and production patterns all over the world. The structural transformation that took place in China thus disrupted the global economy and became the foundation for global political disruptions as well.

China was not alone in transforming the world 'as it was'. Approximately a decade after China began its path towards economic growth, the Indian government introduced a number of reforms that kick-started the Indian economy. Like China, India's size mattered. Although not quite as populous as China, the population of India still makes up approximately 17 per cent of the world's population. Even though structural transformation in China and India have led to increasing wages in these countries, they both possess large reserves of low-wage labour. Structural transformation in these two countries therefore in the short run do not give rise to movement of production to other parts of the Global South on a large scale. In contrast, countries of the Global South that compete on the market for light manufacturing such as Bangladesh and Egypt (textiles and clothing) and services such as Kenya (ICT) will continue to face competition on export markets for a long time (Donno and Rudra 2014).

The growing economic power of countries like China and India did not go unnoticed – neither by the politicians in these 'emerging economies', nor among commentators from the Global North. It first led to a recognition of potential investment possibilities (O'Neill 2001); then to aspirations for political influence among political leaders in the Global South and an aspiration to rebalance global governance; and finally, to an acknowledgement in the Global North that we were witnessing a process of 'shifting wealth' in the world (OECD 2010).

Although at first merely a category developed by the then head of global economic research in Goldman Sachs to attract foreign investments, the acronym BRIC(S) came to signal the changes that took place at the turn of the millennium after two decades of rapid economic growth in China and a decade of growth in India. Likewise, IBSA came to be seen (for a while at least) as an important institution in the turn towards a multipolar world that (also) gives voice to the Global South. It is to these two groups that we turn to now.

The BRIC(S): from an investment category to a political construct

Even if BRIC was originally merely a catchy acronym to point investors towards four economies, i.e. Brazil, Russia, India, and China, that Goldman Sachs predicted would grow so fast that within less than four decades they would catch-up with the leading economies of the Global North, the acronym soon became well-known in international politics and a defining feature of the tectonic shifts in global power relations in the past one and a half decades.

A number of factors account for the unforeseen importance of the term. First, for roughly a decade the four countries grew at a much faster rate than originally predicted (until the financial crisis 'hit hard' on some of them). Second, due to the adverse effects of the global financial crisis on the economies of the Global North the size of these economies either shrank or grew at a much slower rate than expected.

In effect, the convergence between the BRICs and the Global North happened at a much faster speed than projected by Goldman Sachs (Carmody 2013). Finally, a process of internal institutionalisation took place, resulting in the development of BRIC as a political unit.

But what is BRIC(S) really? As stated above, it originated in 2001 as a listing of countries of a particular size (population and GDP-wise) whose economies grew at a particularly (fast) rate. In other words, it was at best financial branding. It was financial branding that worked, though. Only a couple of years after the term had first been coined hardly anyone had to explain it. Alongside its growing popularity, its content changed. In 2001, the foreign ministers of Russia, India, and China initiated a series of annual trilateral meetings focusing on security issues in the region. In 2006, then Russian minister of foreign affairs Sergey Lavrov, proposed to also invite Brazil to these meetings and thus make BRIC into a political group. The first informal meeting, on 20 September 2006, did touch upon security issues but soon revolved around the global governance and in particular the distribution of power in the IMF and the World Bank.

Global governance and reform of international structures were also high on the agenda of the first formal BRIC meeting in Yekaterinburg, Russia in 2008 (for foreign ministers only). Between 2006 and 2008, annual economic growth of the BRIC countries was 10.7 per cent – much higher than in the Global North (China Daily 2009). In 2008, Lehman Brothers went bankrupt and the global financial crisis erupted. This proved a great opportunity to push for changes in global financial governance and it (coincidentally) meant that BRIC cooperation transformed from security issues to international finance.

A year later, in June 2009, the first of a series of yearly BRIC leaders' summits was held. The agenda followed directly from the meeting in 2008 but pushed the demand for changes even further. In particular, BRIC leaders called for greater voice and representation in international financial institutions for emerging and developing countries; an end to the informal agreement between Europe and the US that Europe appoints a president for the IMF and the US appoints a president for the World Bank;[1] and a reduction of the global dependence on the US dollar (Stuenkel 2015a).

The meeting came to signal a break with Goldman Sachs' investment vision. Not only had the group formalised itself in Yekaterinburg, the

Plate 3.1   *South Africa – new boy at the BRICS club*

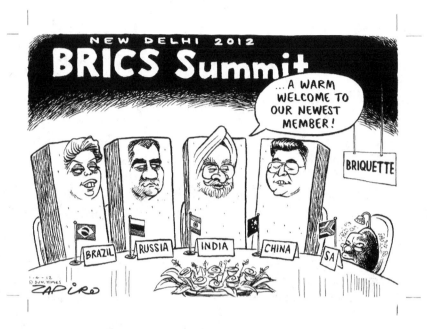

Source: (BRICS Summit © 2012 Zapiro. Originally published in Sunday Times. Re-Published with permission – For more Zapiro cartoons visit www.zapiro.com)

meeting also became the starting point of the enlargement of BRIC to also include South Africa in 2011 and thus change name to BRICS and distance itself from the investment category and develop into a (geo)political formation. According to Stuenkel (2015a), the Minister of Foreign Affairs in South Africa first approached BRIC during the 2009 meeting. In the months that followed South Africa sought simultaneously to strengthen bilateral ties with the four BRIC members and project itself as the regional leader of Africa. By December the following year, South Africa succeeded: it was invited to the third meeting that was to take place in Sanya, China, in April 2011.

The expansion of BRIC to BRICS affected the grouping more broadly. First, BRICS became a true global player. While South Africa's GDP is significantly smaller than for instance China's GDP and its economic growth rates have been trivial compared to China's, South Africa's inclusion in the grouping can been perceived as a gateway to the African continent (Carmody 2013). Second, and relatedly, it boosted BRICS vis-à-vis the Global South: by including an African country, it signalled a change from the approach of most other international institutions. Thus, BRICS came to strengthen SSC. Finally, it further changed the global perception of Africa – away from the 'hopeless continent' (The Economist 2000), characterised by HIV, poverty, and conflicts, towards a continent of hope (The Economist 2011).

India-Brazil-South Africa Dialogue Forum: BRICked up?

Even though IBSA was formed as a political grouping several years before BRIC(S), it developed in the shadows of BRIC(S). IBSA grew out of discontent with global power configurations: when the G8 members[2] met in France in 2003, they invited India, Brazil, South Africa (and China and Mexico) as observers – the so-called 'outreach 5'. None of the invitees, however, were content with this status. They were all middle powers and had for a long time used their regional power status to represent other developing countries, but they were not invited to discuss issues of global importance with the most powerful nations in the world (da Silva, Spohr, and da Silveira 2016).

Hence, on the back of the G8 summit, India, Brazil, South Africa decided to form a new alliance, IBSA. As described in detail by Stuenkel (2015b) the three countries were not well connected in 2003. Trade between them was insignificant – and so were foreign direct investments. Thus, the grouping did not grow out of long-term close economic ties. Rather, it came about as a combination of a reaction to the exclusion of the Global South in international relations; a call for closer cooperation between (democratic) regional powers; and an attempt to increase global political leverage. Immediately after the formation of IBSA the grouping adopted the 'Brasília Declaration', which called for a reorganisation of the world – not unlike NAM and the NIEO; a reform of the United Nations Security Council (UNSC); and inclusive growth and social equity. In order to meet the aims of the Declaration IBSA members were to consult each other and cooperate on global political issues; encourage trilateral cooperation on specific areas via working groups; and assist other developing countries via the IBSA Fund (IBSA Facility for Alleviation of Poverty and Hunger). In contrast to BRICS therefore, which only revolves around high-level summits, IBSA conducts most of its work via a combination of ministerial-level meetings, presidential summits, and working groups.

According to Stuenkel (2015b), IBSA has passed through three phases: the establishment phase (2004–2006); the action phase (2007–2011); and the slowdown phase (2012–2014). The establishment phase was characterised by intense meeting activity, praise and hope, institution building, and contestation over regional leadership. While IBSA at first was only a meeting between foreign ministers, the establishment phase saw the development of several institutions to further the coordination of activities. This phase peaked with the first IBSA summit in Brasilia in 2006 and the IBSA agenda was characterised by the following two issues: advancing South-South Cooperation through preferential trade agreements and cooperation within education, and reform of the UNSC. While extending SSC was relatively uncontroversial, reform of the UNSC was highly controversial. One the one hand, it was opposed by the permanent members and on the other hand, IBSA countries' regional leadership was opposed by their neighbours.

The action phase built directly on the first phase. It was characterised by an ever-increasing number of meetings and actors involved. While business people had been a central part of the vision of advancing SSC they now became an integral part of some of the working groups. The same applies to civil society organisations that were included to some extent to increase ownership of the IBSA idea. Alongside the increase of magnitude and scope of IBSA the leaders formulated a geopolitical vision of multipolarity of world power. The speed of the process, however, was slowed down by the fact that India and Brazil began to step up their engagement in BRIC in 2008, the first BRIC summit in 2009, and the enlargement of BRIC to also include the last IBSA member, South Africa.

BRICS was not the only forum that challenged IBSA. So did G20,[3] and even though IBSA sought to differentiate itself from BRICS by highlighting issues such as democracy and human rights, the period after 2012 is also characterised by slowdown in activities. Due to internal challenges in Brazil and South Africa, the sixth IBSA summit to take place in India in June 2013 was postponed a number of times. Eventually, it was cancelled and instead the three countries managed to meet on the sidelines of the UN General Assembly in New York and celebrate the tenth anniversary of the 'Brasília Declaration'. Thus, 2011 marked the last IBSA summit, and although the sixth IBSA summit officially is still being planned the chance that IBSA will be a driving force in future SSC is relatively slim.

The slowdown, however, did not mean that IBSA did not have an impact. As stated above, intra-BRIC economic interaction was microscopic when the organisation took form at the turn of the millennium. By 2012, intra-IBSA trade had doubled as a share of their total trade. More importantly, their economic engagement with the rest of the world had increased rapidly in the period. While total IBSA imports, for instance, made up only USD 154 billion in 2003 it had grown to USD 814 billion by 2012. Of this only roughly USD 26 billion was made up of intra-IBSA trade (see Table 3.2). By far the majority of the increase in IBSA imports consist of imports from China (Woolfrey 2013). IBSA also aimed to speak with one voice in international relations and engaged fiercely in WTO discussions on trade liberalisation and made sure it had joint positions in security issues.

Finally, IBSA has had some impact via the IBSA Fund – the development cooperation arm of the organisation. The fund was created in 2004 but only became operational in 2006. It aims to fund projects fighting poverty and hunger in collaboration with the UN that are 'replicable and scalable'; to develop best practices; and to advance SSC. The projects take as their starting point that IBSA countries have (recent) experiences in fighting poverty and hunger in their own

Table 3.2 *Intra-IBSA trade, 2003–2012*

	2003	*2006*	*2009*	*2012*	*GROWTH 2003–2012*
TOTAL IBSA IMPORTS (USD MN)	153,842	332,429	450,173	813,953	20.3%
INTRA-IBSA IMPORTS (USD MN)	4,060	8,282	13,663	25,586	22.7%
INTRA-IBSA IMPORTS AS PERCENTAGE OF TOTAL IMPORTS	2.6	2.5	3	3.1	NA
TOTAL IBSA EXPORTS (USD MN)	165,526	316,627	380,576	625,722	15.9%
INTRA-IBSA EXPORTS (USD MN)	2,688	7,176	10,790	22,953	26.9%
INTRA-IBSA EXPORTS AS PERCENTAGE OF TOTAL EXPORTS	1.6	2.3	2.8	3.7	NA

Source: (Woolfrey 2013: 13)

countries that are transferable to other countries in the Global South and the funding comes from both IBSA countries (contributing a minimum of USD 1 mn/year since 2004) and topped up by the UNDP. So far the relatively small-scale projects have focused on the following sectors: agriculture, health, fisheries, renewable energy, water and sanitation, and youth (UNDP 2017; Vieira 2013).

Regional Cooperation fora

Regional cooperation and regionalism mean institutional arrangements to facilitate trade and economic cooperation most often within a spatially defined area. Regionalism may denote formal projects, i.e. the formation of interstate associations/groupings based on regions as well as ideas and values that may or may not lead to institution building, and it may be studied from a number of different approaches – each revealing different aspects of the process (Söderbaum 2004). Regionalism may take many forms, including free trade areas, customs unions, common market, and economic unions. Regional cooperation and regionalism is neither new nor confined to the Global South. Regionalism may denote formal projects, i.e. the formation of interstate associations/groupings based on regions as well as ideas and values that may or may not lead to institution building, and it may be studied from a number of different approaches – each revealing different aspects of the process (Söderbaum 2004).

In fact, regionalism in the Global South has passed several phases. A relatively short-lived first wave of regionalism took off in the 1950s whereupon a second wave of regionalism surfaced in the Global South in the 1980s. In Europe as well as in North America, regionalism took new forms with the establishment of the European Union and the creation of the North American Free Trade Agreement in the 1990s. The Global South began a process of establishing regional bodies to counteract this trend and by the turn of the millennium, a third wave of regionalism in the Global South emerged.

The Bolivarian Alliance for the Peoples of Our America – Peoples' Trade Agreement (ALBA-TCP)

Regionalisation is thus not a new invention – neither in Latin America, nor in the rest of the world. Form and scope of regionalisation, however, seem to reflect dominant politics and ideologies and hence change over time (Muhr 2011). The Bolivarian Alliance for the Peoples of Our America – Peoples' Trade Agreement (ALBA-TCP) is a paradigmatic case of how politics and global governance influence the creation and work of regional entities.

The organisation was founded in 2004 by Cuba and Venezuela as an alternative to the Free Trade Area of the Americas and thus a way to halt it. In contrast to other regional entities, it was articulated as '*a proposal of integral, economic, social, political and cultural integration of the peoples of Latin America and the Caribbean*' (ALBA-TCP 2017), i.e. a new model of regional integration built on new patterns of trade and investment relations between the members (Linares 2011). A couple of years later, in April 2006, the Peoples' Trade Treaty (TCP) was added to the organisational setup. In essence, TCP seeks to make possible exchange between the peoples of ALBA based upon solidarity and complementarity in contract to free trade agreements based on competition.

ALBA-TCP is an intergovernmental organisation currently comprising the following countries: Venezuela, Cuba, Bolivia, Nicaragua, Ecuador, Dominica, Antigua and Barbuda, Saint Vincent and the Grenadines, Saint Kitts and Nevis, and Santa Lucia. It is thus a very heterogeneous unit comprising resource-rich relatively large middle-income economies, socialist economies, poor large countries, and some extremely small island states. What united them – at least when the organisation was founded – was a critique of both the market-led globalisation process and the global power imbalances. This critique is directly observable in the principles upon which the organisation works. For ALBA these principles include: trade must not be an end in itself but only a means to reach sustainable and just development; integration is built on special and differential treatment among the members; non-competition, i.e. specialisation, is built on cooperation and complementarity; the reduction of dependence on external investments by promotion of Latin American investments in the region; and coordination of positions in multilateral spheres such as the UN. For TCP they include: no interference in internal affairs of other member states; the search for complementarity, cooperation and solidarity between the countries so as to advance existing capacities and save resources; protection of national production by only exporting the excesses; imposing rules that favour the smallest economies; development of food security in each member state; and strengthening the role of the state in basic and strategic services such as water, education and health (ALBA-TCP 2017).

While it is relatively easy to understand why ALBA-TCP was founded in the first place (a critique of a free trade agreement specifically and US imperialism more generally), it is more difficult to fully comprehend its role today. On the one hand, the initial backing of the organisation from Latin American left-of-centre countries suggests a demand for a re-politicisation of the region (Emerson 2013). On the other hand, this demand seemed to evaporate a year after the global financial crisis that exposed some of the neoliberal globalisation project's shortcomings. Since June 2009, only three tiny Latin American island states have joined the organisation and since December 2014, no summits have been held.

Likewise, it is unclear exactly how ALBA-TCP wants to pursue the re-politicisation. The principles all share an opposition to neoliberal market-based regionalism, but the organisation is less clear on how to make these principles work in practice. In order to realise them, the organisation introduced two instruments: the 'grannational' project, i.e. projects implemented by two or more ALBA-TCP members in another ALBA-TCP member state following the overall principles, and the 'grannational' enterprise, i.e. companies from member states make direct sales of products to other member states based on principles of fair trade. Even though the organisation envisages 'grannational' projects in a number of sectors including food, water and sanitation, science and technology, education, and energy, none of them seem to have materialised. Similarly, ALBA-TCP imagines 'grannational' enterprises in for instance forest and mining sectors. Again, the main vehicle for making ALBA-TCP work has not commenced.

Hence, even though '*ALBA-TCP epitomises horizontal integration with the declared objective of constructing a more democratic multi-polar world order*' (Muhr 2011: 105), and makes direct reference to the NIEO by focusing on 'cooperative advantage' instead of 'comparative advantage' and detaches itself from the open, deregulated, market-led development strategies of the 1980s and 1990s by referring to endogenous development, it never managed to go beyond the critique of free trade.

South-South Development comes full circle: the Nairobi protocol and the 40 years anniversary of the Buenos Aires Plan of Action

The UN system did not miss the resurgence of SSC, the funds, programmes, and specialised agencies as well as the General Assembly produced reports that sketched the new situation. UNCTAD, for instance, produced both broad reports on investments originating in the Global South (UNCTAD 2006, 2005) and more specialised reports focusing on particular regions (UNCTAD 2010). At the more general level the General Assembly took stock of the current state of SSC as well as evaluating the effect of 30 years of SCC on development (United Nations 2009a, 2009b). Together these reports documented the increasing importance of SSC in terms of trade, investments, and aid.

Partly as a consequence of SSC's growing importance, partly as a way to curb the negative effects on the Global South of the global financial crisis the UN decided to arrange a high-level conference on South-South Development on 1–3 December 2009, in Nairobi. The conference in many ways departed from the Buenos Aires Plan of Action (see chapter 2) adopted three decades earlier. Thus a key theme was technical cooperation between countries of the Global South (assisted by the UN) as a means to further SCC.

The outcome of the conference was published in the Nairobi Resolution (64/222). Most importantly, the resolution reaffirmed the conclusions from the Buenos Aires Plan of Action – focusing in particular on the 'proximity of experience' of the actors of the Global South that was perceived to further capacity building; it acknowledged the heterogeneity of actors of the Global South; it recognised that SSC does not substitute but complement North-South Cooperation; and appreciated that SSC is not development aid, but rather a '*partnership among equals built on solidarity*'. Moreover, it gave the UN (including the programmes and the specialised agencies) a special role in promoting and coordinating SSC and triangular cooperation.

The UN General Assembly is currently planning to celebrate the 40 years anniversary of the Buenos Aires Plan of Action in Buenos Aires, Argentina, from 20–22 March 2019.

Conclusion

To a large extent, BRICS and IBSA continue where NAM and NIEO ended (see chapter 2). Both institutions are (partly) rooted in the Global South; both institutions seek to alter global power relations and give more voice to the countries of the Global South, but in contrast to the NIEO that was rooted in structuralist economics and called for government intervention and protectionism (see also Textbox 2.2), IBSA argues that liberalism has not gone far enough as trade protectionism benefits the Global North vis-à-vis the Global South. Even though BRICS and IBSA therefore have the same overall aims, they also differ in many ways. First and foremost, IBSA has called for a reform of UNSC whereas BRICS so far has refrained from debating this issue. Second, IBSA members are all multiparty democracies while BRICS also includes authoritarian regimes. Third, their economic power differs radically: while BRICS has established the New Development Bank and the Contingent Reserve Arrangement (see chapter 4) to back up its aim to further development at home and abroad, IBSA has established the much smaller IBSA Fund focusing solely on alleviating poverty and hunger.

Discussion questions

- Explain why the rise of China and India mattered more for the Global North than the rise of first and second-tier NICs.
- Describe the differences between BRICS and IBSA and discuss the extent to which their objectives conflict.
- Explain how BRICS transformed from an investment category to a political construct and consider what this will mean for BRICS in the future.
- Explain how BRICS differs from ALBA-TCP and discuss which group will impact SSC most in the future.

Web pages of interest

- The official India-Brazil-South Africa Dialogue Forum (IBSA) website is apparently no longer updated regularly, but it contains information from the first five IBSA summits: www.ibsa-trilateral. org/#

- BRICS 2017 is the official website of the most recent BRICS summit held in Xianmen in September 2017. It includes all official documents from the summit: www.brics2017.org/English
- The ALBA portal is the official website for the Bolivarian Alliance for the Peoples of Our America. It offers rich information (in Spanish) on issues related to the members including a list of novels of particular relevance to understanding the aim of the organisation: www.portalalba.org
- Network of Southern Think Tanks aims to provide a platform for think tanks in the Global South to generate and share knowledge on SSC. The web site contains policy briefs on SSC seen from a Global South perspective: www.southernthinktanks.org
- The South-South Conference website is the official web page for the high-level conference on South-South Cooperation held in Nairobi 2009. It contains all the official documents as well as the programme: www.southsouthconference.org
- South Centre is an intergovernmental organisation established in 1995 to replace the South Commission. The centre undertakes research and analyses of issues of importance for the Global South: www.southcentre.int

Notes

1 Although the members are formally aligned concerning leadership of for instance the IMF, practice sometimes turns out a bit more blurred. When former managing director of the IMF, Dominic Strauss Kahn, stepped down in 2011, BRICS members did not agree on his replacement and Brazil and China supported the candidate preferred by the US, the French Christine Lagarde. A year later, when Robert Zoellick's position as President of the World Bank was ending, BRICS countries also split in their support for his successor (Gray and Gills 2016).

2 G8 was an intergovernmental forum in existence from 1997 to 2014. It brought together representatives from France, Germany, Italy, Japan, Great Britain, the United States, Canada, and Russia. In 2014, after the annexation of Crimea, Russia was suspended from the G8 and the grouping renamed itself to G7.

3 The G20 is an international forum focusing in particular on issues of global economic governance. It is made up of 19 individual countries and the European Union thus representing some 85 per

cent of world GDP, 66 per cent of world population, and 75% of international trade.

Further reading

Carmody, P. 2013. *The rise of the BRICS in Africa. The geopolitics of South-South relations*. London: Zed Books. A lucid book on how individual BRICS countries affect African countries. Based on five independent studies Carmody, offers a very accessible analysis of how globalistion affects Africa and how and to what extent it matters for Africa's place in international relations.

Farooki, M., and R. Kaplinsky. 2012. *The Impact of China on Global Commodity Prices. The global reshaping of the resource sector*. New York: Routledge. A detailed analysis of how resources affect development and how China's (and to some extent India's) growing demand for a variety of commodities has affected the rest of the Global South.

Leftwich, A. 1995. Bringing politics back in: Towards a model of the developmental state. *Journal of Development Studies* 31 (3):400–427. This article describes the developments that took place in the 'third world' from the 1960s to the 1990s and argues that the best way to understand this development is through the analytical lens of the so-called developmental states. It outlines the characteristics of developmental states and thereby brings politics back into the fireld of development studies.

Stuenkel, O. 2015b. *India-Brazil-South Africa dialogue forum (IBSA): the rise of the global south?* London: Routledge. One of the few books on IBSA. It offers a detailed description of how it was established and how it developed over time.

4

Vectors of South-South Cooperation

Aid: catalysing the growing engagement

In her very comprehensive book about emerging donors and the changing development landscape Mawdsley (2012b) concludes that no matter which international relations theory perspective one adopts, aid essentially is political. It is a foreign policy tool used to advance certain strategic and/or economic ambitions. However, this does not entail that all donor ideas and practices are identical. In contrast, donor ideas and practices are shaped both by diplomacy and domestic politics in the donor country (Lancaster 2008) and, for emerging donors, their histories as aid recipients (Kragelund 2011). Likewise, the precise amount, focus (sectoral and country-wise), and modality of aid is affected by geopolitics, international events, and different 'fashions'. Hence, purposes such as commerce, democracy, and global public goods have been given different attention in national aid policies at different epochs.

Box 4.1 Facilitating trade and investments: Indian aid to the Pacific Island Countries

Most accounts of South-South aid examine Chinese development finance to sub-Saharan Africa. True, there is now a growing interest in Chinese interest in Latin America, the Middle East, and the rest of Asia. Likewise, the interest in Brazilian, Indian, and Middle Eastern aid to Africa is also growing, but beyond this SSC is unchartered territory. This is not tantamount to lack of activity. Growing Indian interest in the Pacific Island countries is exactly an example of activity that has not been thoroughly documented yet.

India provides aid[1] to more than 160 countries around the world and although all this aid is cast in the same SSC terms, the size and modalities of India's aid depends on the perceived geopolitical importance of the recipient. Even though Indian diaspora makes up more than one-third of the population in Fiji, for instance, India has only disbursed aid to the country for roughly a decade – and only after China began to expand its activities in the region. Since then, however, India has boosted its activities in countries like Fiji and Papua New Guinea.

Driven by commercial and political competition with China, India has passed an 'Act East' policy that included (re)opening High Commissions, upscaling aid activities, introducing credit lines, and expanding the scholarship programmes. India now uses the full aid palette to grow ties with the Indian diaspora, increase trade, access natural resources, and get support in international fora.

Source: (Zhang and Shivakumar 2017)

In this sense, South-South aid does not differ radically from North-South aid. Aid disbursement in the Global South is used to facilitate political, strategic, economic (and humanitarian) objectives at home as well as abroad (see Textbox 4.1). Within both South-South and North-South aid, however, heterogeneity is massive, and focus, modality, and magnitude change over time. Schematically, however, what distinguishes North-South from South-South aid flows is 1) their acceptance of DAC principles; 2) the degree to which all flows are captured by one organisation governing 'aid' and hence the degree to which national statistics reflect the flows; 3) the importance they ascribe to multilateral and bilateral aid, respectively; 4) the extent to which donors collaborate to reach their aim; 5) the degree to which they are open about the strategic, economic, and political purposes of aid; 6) the use of conditionalities to change policies in recipient countries; 7) who they engage with locally; and 8) the degree to which aid catalyses other types of cross-national engagements (see also Table 4.1).

As spelled out in Textbox 1.3, the DAC specifies what aid is for the Global North and how it is governed. The Global South has no comparable organisation defining aid. Instead, each and every Southern donor applies its own definition of aid. Some of these donors' aid reflect many of the DAC principles either because they seek assistance from the DAC to organise their aid or because they have accepted an invitation to become an associate of the DAC (can take part in all meetings and decision-making processes of the DAC) or a participant to the DAC (can take part in non-confidential meetings).[2] Other Southern donors do not apply DAC standards and definitions and, hence, comparing their 'aid' to ODA is more difficult. However, if one adopts a broader definition of 'development finance' that also includes 'other official flows'

Table 4.1 *Schematic overview of North-South vs. South-South development finance*

	SOUTH-SOUTH	*NORTH-SOUTH*
ACCEPTANCE OF DAC PRINCIPLES	ODA and OOF are mixed	Clear separation between ODA and OOF
GOVERNANCE STRUCTURE	Overlapping, competing institutions and embryonic strategies	Governed by strategies, policies and clear institutional framework
BILATERAL VS. MULTILATERAL CHANNELS	Primarily bilateral channels	Both multilateral and bilateral channels
DONOR COLLABORATION	Occasional trilateral cooperation	Common donor meetings
OFFICIAL PURPOSE	Geopolitics and economic development at home and abroad	Social and economic development in recipient countries
USE OF CONDITIONALITY	'No strings attached'	Widespread use of *ex ante* and *ex post* economic and political conditonalities
STATE VS. CIVIL SOCIETY	Respect for national sovereignty	Involvement of civil society
DEGREE OF CONNECTION TO OTHER FLOWS	Development finance facilitates trade and investments	ODA separated from trade and investments

(OOF), i.e. transactions from the official sector that do not meet ODA criteria, for instance a grant element below 25 per cent and/or transactions aimed primarily to enhance exports, it becomes less troublesome to compare South-South aid with North-South aid (Bräutigam 2011). The ODA-like part of the transactions then comprise central government foreign aid expenditures (grants and interest-free loans), debt relief, scholarships, and the value of interest subsidies of concessional loans. On top of this comes OOF-like transactions like export credits and subsidies to private sector actors to ease their investment in the Global South.

Another major difference between South-South aid and North-South aid is that the former is seldom governed by one institution only. In contrast, numerous government entities at central and decentral level are involved in distributing development finance (see also chapter 5). Stated differently, North-South aid is characterised by clear strategies, policies, and institutional frameworks for development cooperation whereas South-South flows are typified by lack of clear policies and competing institutions.

Most South-South aid goes through bilateral channels whereas some North-South aid flows, at least historically, were somewhat evenly distributed between bi- and multilateral channels. Of late, however, aid from the Global North increasingly is directed to the Global South via bilateral channels.[3] The use of bilateral channels, i.e. transactions undertaken by a donor country directly with a country in the Global South, transactions with a development NGO, or debt relief, reflects the geostrategic nature of aid. Unlike multilateral aid, bilateral aid can be tracked back to the original donor and it is possible for the donor to pre-define the exact purpose of the transaction. The use of these two channels, respectively, is linked to the degree of openness re. geopolitical and strategic purposes. While the DAC specifies that ODA first and foremost has to be developmental, most donors of the Global South openly admit that aid is also given for strategic purposes: as both donor and recipient lack economic resources, aid has to benefit both partners. It should be noted, though, that even if the DAC requires that the purpose of ODA first and foremost is social and economic development in recipient countries, history is full of examples of 'Northern' donors giving aid for geostrategic, domestic and/or economic reasons (Lancaster 2008).

Linked hereto, donors from the Global North make a virtue of harmonising their approach via common donor meetings. In contrast, Southern donors engage bilaterally with recipient governments and only seldom do they engage in so-called trilateral development cooperation, i.e. a development relationship in which a 'Northern' donor – either bi- or multilateral – teams up with a so-called 'emerging' donor to provide aid in a country in the 'rest of the Global South', see Textbox 6.1.

Donors from the Global North make widespread use of conditionalities to change the economic and political direction of recipient countries. Conditionalities can be applied both *ex post* and *ex ante*. Historically, most conditionalities have been applied *ex post*, i.e. the recipient receives a certain amount of money after having agreed to implement certain reforms after the agreement has been signed. Of late, ever more donors introduce *ex ante* conditionalities, i.e. countries only receive funds after they have demonstrated that they are on the 'right' track. The most well-known conditionalities are linked to the structural adjustment programmes that took off in the 1980s on the back of the debt crisis and the demand for political liberalisation after the end of the Cold War. Donors from the Global South seldom make use of policy conditionalities. Instead, they call for a more pluralistic development, partner-led path where various combinations of degrees of economic and political liberalisation may lead to development.

Directly linked to the use of conditionalities, Southern donors typically used to deal only with the 'President's office' in the recipient country, arguing that the head of state is responsible for directing the country towards development. In contrast, aid from the Global North is characterised by also engaging with civil society – in the North as well as in the South. Southern donors are now gradually changing their strategies including engagement with line ministries as well as non-state partners.

Finally, the DAC actively seeks to detach aid from other flows of finance from North to South by defining the purpose of aid, monitoring the degree of tiedness of donor aid, and describing in detail what can be counted as ODA and what cannot. Changes are currently taking place at the heart of international development cooperation. Internal voices in the DAC have proposed replacing the ODA with a new Official Development Effort measure, which in short returns to the core intention of the ODA, i.e. social and economic development in developing countries. Thereby, it excludes all the 'transfers' now included in the ODA that stay in donor countries, and instead only counts either grants or concessional elements of loans for development purposes in developing countries (Hynes and Scott 2013). Simultaneously, bilateral DAC donors are gradually rejuvenating their focus on economic growth; some development agencies are moving back to ministries of foreign affairs or ministries of trade; and focus is increasingly on mutual benefits (Mawdsley 2015). In contrast, Southern donors actively link aid to other vectors of engagement. Chinese development finance, for instance, is used to get access to resources and form the hearts and minds of the next generation of politicians in the Global South. Similarly, United Arab Emirates (UAE) aid to Egypt is used to advance its own investment in the country, see Textbox 4.2.

Box 4.2 United Arab Emirates aid to Egypt

Some Gulf states, i.e. UAE, Kuwait, Qatar, and Saudi Arabia, have a relatively long history of providing aid – at first only to other Arab and/or Muslim countries but since the turn of the century also to non-Arab countries.[4] UAE's aid programme began with the establishment of the Federation in 1971. It built on Abu Dhabi aid (the Abu Dhabi Fund for Development) experiences that date back to 1969 as well as on the experiences of the OPEC Fund for Development. Despite its long history of aid provision large gaps exist in our knowledge of the scope and magnitude of this aid as its reporting has been inconsistent.

Since then UAE aid has come a long way. On the one hand, it has become more formalised and more aligned to DAC principles. In 2013, for instance, UAE formed a Ministry of International Cooperation to manage its aid programme; cooperate with other donors; and better respond to humanitarian crises. In 2016, it was merged with the Ministry of Foreign Affairs and launched under its new name, the Ministry of Foreign Affairs and International Cooperation. Furthermore, since 2014, UAE has been a 'participant' of the DAC. Participants are not members of the DAC but can '*contribute to discussions on key development issues and benefit from DAC members' experiences*' and '*can take part in non-confidential meetings of the DAC*' (OECD ND-a). Thereby, a participant may learn and share experiences, but does not take decisions. Alongside this new DAC status, UAE has begun to publish its aid giving to the DAC using DAC definitions. UAE has also begun to publish its aid flows using DAC criteria on the DAC website.

On the other hand, UAE, alongside other Gulf donors, has recently developed a set of new aid modalities that diverge radically from DAC aid. These modalities include non-restricted cash grants, in-kind oil and gas deliveries, and financial injections into recipients' central banks: they clearly mix aid with investments.

The case of UAE aid to Egypt provides a good illustration of how these conflicting tendencies play out in reality. UAE has economic, ideological, and security interests in Egypt: Emeriti state-owned firms have economic interests in the banking, retail, construction, and agricultural sectors in Egypt; UAE supports the secularism of President Abdel Fata al-Sisi of Egypt (against the Muslim Brotherhood) and needs Egyptian support for its military operations in Yemen. In contrast, Egypt desperately needs money: investors stay out mainly due to domestic security issues and the President needs cheap oil and gas to keep the population from protesting openly. This mutual interest has led to a large increase in transfers of development finance from UAE to Egypt lately.

According to Young (2017: Table 1), UAE's development finance to Egypt, 2011–2016, includes: USD 3 billion for construction/real estate and development of small- and medium-sized enterprises (2011); USD 22.8 million in private aid (2011); USD 22.19 million in private aid (2012); a USD 3 billion injection into the Central Bank of Egypt (2013); in-kind oil and gas deliveries worth USD 225 million (2013); a USD 2 billion injection into the Central Bank of Egypt (2015); a pledge of USD 2 billion to revive the Egyptian economy (2015); a USD 2 billion injection into the Central Bank of Egypt (2016); and a pledge of USD 2 billion worth of investments to revive the Egyptian economy (2016).

Obviously, not all of this is comparable to ODA. This is also evident if we compare the figures above with what is recorded by the DAC. According to DAC, UAE transferred USD 2.46 billion to Egypt in 2015 (OECD ND-b). The difference from the DAC aid is immediately discernible. UAE mixes aid with investments (and trade); it uses aid to export its own political economy models (including boosting central banks to depreciate the currency); it uses aid to advance its own investments; it deals only with the President's office; it makes widespread use of pledges; it is unpredictable; and nearly all (99 per cent) aid flows are bilateral. All of these characteristics are in sharp contrast to the policies of the DAC (see Textbox 1.3).

Sources: (Young 2017; Shushan and Marcoux 2011; Momani and Ennis 2012; Al-Mezaini 2017; OECD ND-b)

Humanitarian assistance

Modern humanitarianism is often traced back to the 1859 battle of Solferino, Italy, when Franco-Sardinian troops fought Austrian soldiers. Some 6,000 men died and another 35,000 men were wounded. The medical services of the fighting armies were not the least bit prepared to help the wounded and dying. Instead, Henry Dunant, a businessman who was in the area, did his best to assist them. This episode first led to a call for the establishment of a neutral and impartial organisation, the International Committee of the Red Cross, to help the wounded and dying and then to an agreement on the core principles of humanitarianism, namely humanity, impartiality, neutrality, and independence.

Although humanitarianism is international by nature, it has often been regarded as a North-South endeavour (Pacitto and Fiddian-Qasmiyeh 2013). The Global North has thus defined the norms and perceived humanitarianism as '*activities designed to save lives, alleviate suffering and maintain and protect human dignity during and in the aftermath of emergencies*' (Binder and Meier 2011: 1137). This, to some extent, is in contrast with Southern actors who perceive humanitarianism to encompass '*all forms of selfless help to people in need, including religious charity, development co-operation, and assistance in times of disaster*' (Binder and Meier 2011: 1137). In fact, however, the guiding principles are almost similar: South-South humanitarianism is officially guided by impartiality, neutrality, and respect of sovereignty.

This leads (most often) to the avoidance of assistance related to conflicts and instead a focus on assistance related to natural disasters and/or epidemics, but it does not entail that South-South humanitarianism is apolitical. Like development finance more broadly, South-South humanitarianism is both an international soft power tool as well as a domestic political tool. It is a means to portray the political power of the Southern donor internationally; it demonstrates responsible statehood; it is a way of dispelling neighbours' potential fear of growing economic and political power; and a way to show migrants and refugees from areas affected by humanitarian crises that the government of their new (temporary) home cares.

The growth of South-South humanitarianism is noticeable. According to O'Hagan and Hirono (2014: 412), humanitarian assistance from the non-Western donors rose almost 18 times in ten years: from 2000 (USD 35 million) to 2010 (623 million). Most attention has been given to donors from the Emerging South but the phenomenon is by no means confined to these countries. The rest of the Global South is also contributing to South-South humanitarianism (Pacitto and Fiddian-Qasmiyeh 2013). Financial Tracking Service has since 1992 tracked donors' financial reporting of humanitarian assistance. According to their database, Benin, for instance, contributed approximately USD 150 million to Haiti in 2010; Ecuador donated USD 800,000 mostly to emergency shelter in 2014; the Philippines granted the Central Emergency Response Fund more than USD 6 million in 2012; Tanzania provided USD 233,000 to Somalia via the World Food Programme in 2012; and Zambia donated USD 80,000 to Kenya via the World Food Programme in 2012 (Financial Tracking Service 2018). On top of this come material contributions, in-kind contributions, and contributions from civil society in the Global South. What we see now, therefore, is a diverse set of actors from the Global South donating finances and materials to other countries in the Global South affected by humanitarian crises.

On the one hand, this could be perceived as a democratisation of humanitarian assistance leading to new insights and new modalities that eventually could benefit the intended beneficiaries. On the other hand, concerns have been raised regarding the fragmentation of assistance. This came to the fore, in particular, after the 2010 Haiti earthquake when more than 100 states donated money. In the words of Kot-Majewska (2015: 122):

> *misunderstanding of each other's values or intentions while engaging in humanitarian action can lead to duplications,*

> *gaps, or inefficiencies in response, which in turn create risks*
> *for the perception of humanitarian actors and their principled*
> *approach, not to mention negative consequences for the*
> *beneficiaries.*

Trade: booming, but structurally unequal

The number of reports that have concluded that the world is
transforming has been booming in the past decade. Many of these
reports revolve around changing trade patterns, i.e. the extent to which
world exports originate in the Global South, the relative size of South-
South trade compared to, for instance, North-North trade, the degree
to which economic growth in the Global South is based on trade.
The figures no doubt point in one direction – and one direction only
(see Table 4.2) – namely that trade patterns are shifting: almost half
of world exports originate in the Global South; and in less than two
decades South-South trade more than doubled its share of world trade
(and the relative importance of North-North trade has declined as a
consequence).

These trends come as no surprise as the economic growth of countries
like China and India since the 1980s and 1990s, respectively, has led
to a shift in global demand (see also chapter 3): the Global South is no
longer only the home of exports of primary commodities; the lion's
share of consumer goods are now produced in the Global South. This
has led to a shift in the distribution of value addition in manufacturing

Table 4.2   *Shifting geography of trade*

	THEN (YEAR)	*NOW (2012)*
SOUTH % OF GLOBAL GDP[a]	21.7 (1980)	35.8
EXPORT AS % OF GDP FOR SOUTHERN COUNTRIES[a]	16.7 (1981)	29.5
SOUTH AS % OF WORLD EXPORTS[b]	29.6 (1980)	44.7
SOUTH-SOUTH AS % OF GLOBAL TRADE[b]	11.7 (1995)	25.5
NORTH-NORTH AS % OF GLOBAL TRADE[b]	51.2 (1995)	33.9

[a] South defined based on World Bank classifications;
[b] South corresponds to UNCTAD's definition of developing countries.

Source: Horner (2016: Table 1)

to also include value addition in the South and thus an increasing share of world income (OECD 2010). Alongside this change, the population of the Global South is now increasingly demanding light consumer goods and durables that are produced within the Global South. The result is a growing importance of South-South trade.

These aggregate trends, however, do not further our understanding of who is benefitting from this shift, how and why. In order to unpack these aggregate figures, it is important first to examine the flows critically. No doubt, for instance, a large portion of the increase in South-South trade over the past two decades can be traced back to the shifting patterns of production, i.e. most of what is traded in the Global South are intermediate goods used for production of branded manufactured goods ending up for the most part in the Global North. Likewise, changing tariff and non-tariff barriers (most importantly the end of the Multi-Fibre Agreement (MFA) and the establishment of the African Growth and Opportunity Act (AGOA)) have affected production and hence (intra-firm) trade patterns within the Global South.

Since the beginning of the 1970s, the now-expired MFA has governed trade in apparel and textiles via quotas by ratifying developed countries' rights to impose quotas on textiles and clothing imports from developing countries. Specifically, the agreement stipulated how much apparel and textiles a particular country in the Global South could export to a country in the Global North. As not all countries in the Global South were affected by the MFA, it led to a process of 'triangle manufacturing' where buyers in the Global North continued to place their orders with their usual manufacturer in the Global South but as this manufacturer was covered by the changing rules, they shifted their production to subsidiaries in other countries in the Global South not covered by the rules. Hong Kong garment producers, for instance, shifted their production to *inter alia* Mauritius (and later Madagascar) and Taiwanese producers shifted production to among others Lesotho, Swaziland, and South Africa. Thereby, the usual manufacturers in the Global South turned into middle-men, the buyers in the North escaped the quotas, and new centres of apparel and textile production emerged (Gereffi 1999). When the MFA eventually was phased out on 31 December 2004, it led to the relocation of production once again – away from the new centres and back to the usual manufacturers.

AGOA was passed in the US in 2000. Essentially, it aims to assist sub-Saharan African countries to grow economically by further opening their economies and liberalising their political climate. This is done by awarding quota- and duty-free status to among other things clothing articles directly imported into the United States from countries covered by the Act, i.e. countries fulfilling certain economic and political criteria set by the US. AGOA builds on an ingenious system of rules of origin, i.e. a system stipulating which processes must be undertaken locally in order for a product to be considered as local for the exporting country. These rules ensure that value addition takes place in the country that benefits from the regulatory framework. Like MFA, AGOA has had a major impact on the relocation of the global textile and apparel industry in the Global South even if the impact has now faded somewhat away (Gibbon 2003; Phelps, Stillwell, and Wanjiru 2009).

If we are to understand global trade patterns we have to understand both how firms are governed – and thus why they outsource some parts of the production and not others (Gibbon and Ponte 2005) – and how trade is governed, but if we are to assess how these changes affect individual countries/producers we have to dig deeper. First and foremost, it is important to assess whether or to what extent South-South trade makes it easier for firms based in the Global South to upgrade products or processes; to internationalise due to lower standards in the end market; to move into new activities with higher value added; and to minimise risks by diversifying end markets, than South-North trade. Equally importantly we have to assess the degree to which South-South trade is a race to the bottom where competition is even stiffer than South-North trade due to, for instance, lower standards, lower prices, and more firms in the same market (Horner 2016). Empirically we see both types of effects – both an easier path towards upgrading and stiffer competition depending on the specific characteristics of the product, the governance of the chain, and the regulation of the market.

The case of 'greater Chinese' textile and apparel manufacturers' transformation from their original role as 'middle-men' in triangular manufacturing to become what Azmeh and Nadvi (2014) call 'strategic partners' working closely with the buyers in the Global North in areas such as research, design and branding, and coordinating the value chains demonstrates functional upgrading among the so-called 'first-tier suppliers' in the Global South: they no longer just forward orders to producers in countries not affected by schemes such as MFA as they did prior to 2004. Instead, they actively engage in the strategic decisions and manage production across the Global South. Thereby, they are key to understanding the extent to which other Southern producers may benefit or suffer from South-South trade.

Garment production in Jordan illustrates some of these issues. According to Azmeh and Nadvi (2014) garment production was almost non-existent in Jordan in 1997 when the US signed a preferential market agreement (duty and quota-free) with Jordan and Israel with highly flexible rules of origin. On their side, the government in Jordan passed a law to allow for great flows of migrant labourers to work in the garment factories. First-tier suppliers, i.e. companies supplying directly to the mother company, from among others Taiwan, Hong Kong, and Pakistan made use of this access to the US market and literally established a garment industry in the country. A decade later garment exports from Jordan had risen from USD 3 million in 1997 to USD 1.25 billion in 2006. The highly flexible rules of origin and the revised immigration laws compensated for the relatively high labour costs and the lack of experience in garment production in Jordan but it also meant that, for instance, yarns and fabrics were imported into Jordan and that most of the workers in the factories were Asian migrants. The result is that the local learning process is circumvented while simultaneously entry and exit costs for the first-tier supplier are reduced.

The picture that emerges from the South-South garment trade in Jordan is mixed. On the one hand, it reveals the rapid upgrading process that has taken place in many 'Greater Chinese' textile and apparel manufacturers over the past couple of decades and shows how changes in the regulatory framework affect firm-level strategies as well as the geography of global production. On the other hand, the situation for second- and third-tier suppliers, i.e. companies supplying to first- and second-tier firms, respectively, do not seem to have improved a lot due to the integration in value chains coordinated by Southern firms: hardly any knowledge is transferred and production in a particular country is dependent only on the availability of low-cost labour and easy access to end markets (Azmeh and Nadvi 2013).

Investments

In 2006, UNCTAD chose to focus on investments originating in the Global South in its flagship report, the World Investment Report. Four chapters, or more than 140 pages, were devoted to '*FDI from developing and transition economies*' (UNCTAD 2006). While admitting that this was not the first time that the world experienced so-called 'Third World Multinationals' (see also chapter 2), UNCTAD made a case that halfway into the first decade of the new millennium the phenomenon was once again unfolding at a noteworthy speed; that ever more countries from the Global South were taking part in this trend; that these firms covered basically all sectors of the economy; and that in contrast to earlier eras of 'Third world multinationals', some of these new companies were global players. The global financial crisis further incited this trend as the Global South was less affected than the Global North – even if the effects were varied across both South and North. Moreover, it meant that firms from the Global South increasingly targeted markets in other parts of the Global South.

This trend is also depicted in Amighini and Sanfilippo's (2014) analysis of South-South FDI in African economies. They show that from 2003 to 2010 FDI from the Global South made up between 18.4 and 29.3 per cent of all greenfield[5] investments in Africa. A growing share of the greenfield projects derived from the BRIC countries (between 22.2 and 44.8 per cent) and African countries (between 6.9 and 48.7 per cent) and in 2010 FDI from BRIC and Africa into Africa made up approximately 72 per cent of all South-South greenfield FDI projects into African economies (see Figure 4.1).

Compared to North-South greenfield FDI projects in Africa, South-South projects focused more on services (approximately 69 per cent of all projects compared to 60 per cent of North-South projects) and less on manufacturing (24 per cent compared to 30 per cent) and mining (7 per cent compared to 10 per cent) (Amighini and Sanfilippo 2014: Figure 4). These figures, unfortunately, do not inform us about the size of greenfield investments coming into each sector. Neither do they take into account purchases of existing facilities. As shown in Kragelund and Carmody (2016), the global financial crisis led to the change of ownership of a number of mining corporations in Zambia – away from the Global North to multinational corporations (MNCs) originating in the Global South.

Figure 4.1 *South-South greenfield FDI projects into Africa, 2003–2010*

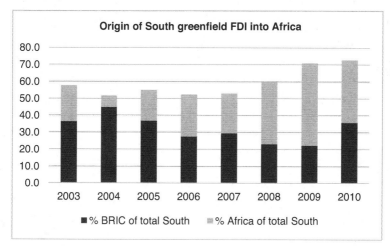

Source: (Amighini and Sanfilippo 2014: Table 1).

Box 4.3 Malaysian FDI in Vietnam

Aggregate data on South-South FDI tends to even out the great differences that exist among Southern investors as well as the changes that take place over time. Malaysia is an illustration of the changes that take place. For a couple of decades Malaysia, as a second-tier NIC (see Textbox 3.1), was a preferred destination for Chinese (and Northern) FDI, but during the past decade new layers have been added to this picture. Since 2007, Malaysia has, like many of its East Asian neighbours, become a net exporter of FDI but compared to some of these economies Malaysia does not possess as strong a manufacturing sector and hence the lion's share of Malaysia's outward FDI is concentrated in the services and agricultural sectors.

A considerable part of these investments target Singapore, Indonesia, Thailand, and Vietnam. While Malaysian FDI into the former three countries has a long history, Malaysian FDI targeting Vietnam is of recent origin. The Malaysian firms investing in Vietnam encompass so-called government-linked companies as well as numerous small- and medium-sized companies.

Based on records of Malaysian firms in Vietnam provided by the Malaysian Business Chamber Vietnam, Lim (2017) interviewed 56 firms to determine why they invested in Vietnam and how this FDI affected the Vietnamese economy. The results show that approximately 70 per cent of the investments were market-seeking, i.e. they targeted Vietnam's 'populous and young domestic market', some 54 per cent were resource-seeking (labour and raw materials), only 34 per cent – mostly in the construction sector – were strategic asset seeking investments, and only 9 per cent were efficiency seeking investments pointing to the comparative low levels of technological capabilities in Malaysia.[6]

The wider effects of the investments are still somewhat ambiguous, but the combination of labour-intensive companies and a relatively low technology gap caters for positive spill-overs. Moreover, Malaysian firms' experience of working in 'opaque' institutional settings ease their operations in Vietnam and thus increase their chances of success relative to firms originating in the Global North.

Source: (Lim 2017).

This new wave of investments from the Global South, often led by emerging multinational corporations (EMNC), has been spurred by processes of globalisation, increasing competition, and liberalisation. This, however, does not entail that EMNCs form a homogeneous unit. In contrast, they differ internally in relation to the size of the home economy, the industrial structure, and support of the government, as well as in relation to the motives and generic strategies to internationalise (Dunning, Kim, and Park 2008). Moreover, they set themselves apart from MNCs from the Global North due to the home country and industry context in which they operate. In particular, the following aspects set EMNCs apart from MNCs: the role of government in decision-making, the role of networks, the characteristics of the markets, and the institutional environment (Gammeltoft, Barnard, and Madhok 2010).

Box 4.4 Exporting China's success to the Global South via 'trade and economic cooperation zones'

Since the mid-1990s China has authorised the establishment of a number of trade and economic cooperation zones (generally known as Special Economic Zones) overseas. Essentially, a Special Economic Zone (SEZ) is a spatially delimited area governed by special (liberal) rules to attract foreign investments. They are built and operated by Chinese companies who negotiate the legal and regulatory framework directly with the host country. Although the governance structures in this manner by definition vary from zone to zone, the rules inside the zones are generally characterised by lower taxes (including VAT rebates), low trade barriers, and low labour standards compared to the rules outside the SEZs. Moreover, companies that invest inside the SEZ are normally allowed to repatriate a larger share of their profits than companies located outside the SEZ.

These SEZs fulfil a number of purposes, including increasing the demand for Chinese goods and services; minimising tariff and non-tariff barriers for trade; facilitating upgrading of Chinese firms; creating economies of scale; and transferring a key ingredient of China's structural transformation to other countries of the Global South. The former four result in generous (but performance-based) grants and subsidised loans from the Chinese state (and provincial governments) to companies entering into the zones, and the latter is why the SEZs are perceived to be of interest for SSC. In theory, the SEZs have the potential to create employment and stimulate manufacturing production in the host economy; to build backward linkages to local companies and thus form the base for knowledge and technology transfer; to build local skills via on-the-job training; and to facilitate improvement of key infrastructure. This is largely what happened when China first experimented with SEZs within China and this is also to a large extent what happened in large parts of East Asia. The question is whether large technology gaps between Chinese firms and local firms' low absorption capacity coupled with political and sectoral constraints make structural transformation via SEZs unachievable in Africa.

So far, government funded/initiated SEZs are up and running in five African countries, namely Algeria, Egypt, Ethiopia, Nigeria, and Zambia. According to Bräutigam and Tang (2014), developments in the African SEZs have been mixed. The Zambian zone has been closely related to one of the Chinese-owned mines, the Chambishi mine, and although the intention has been to locate a larger share of the value-added part of copper production within Zambia, it has been a slow process and the companies located in the zone have not linked up in any great way to locally-owned firms (Kragelund and Carmody 2016; Fessehaie and Morris 2013). The Ethiopian zone is also characterised by lack of linkages to local firms. Although the focus of the zone is light manufacturing (generally known for its plethora of forward and backward linkages) the lack of sectoral focus of the zone has resulted more in access to cheap labour, inputs, and land than in structural transformation for Ethiopia (Giannecchini and Taylor 2018). Up top of this, plans for a SEZ in Mauritius have been underway for a long time. However, the effects of this zone are still to been seen. Due to problems related to resettlement of farmers owning the land where the zone was to be located and economic recession related to the global financial crisis, the zone is still not launched (Bräutigam and Tang 2014).

Two zones have been established in Nigeria: one close to the international airport of Lagos and the other, a bit further away from the city. While the former now accommodates firms from China, India, Lebanon, and Nigeria (though on a much smaller scale than anticipated in 2004 when the project was on the drawing board), the latter is still under construction. The final zone is the one located just south of the Suez Canal in Egypt. It is also the oldest. Although it was initiated almost two decades ago, it still only accommodates approximately 40 firms.

Sources: (Bräutigam and Tang 2014; Giannecchini and Taylor 2018)

Migration

By a rough estimate, in 2005 two of every five migrants on the globe – some 78 million out of 191 million migrants – were residing in a developing country. Most of these migrants are likely to have come from other developing countries.

(Ratha and Shaw 2007: 1)

This is how a World Bank study on South-South migration begins. Since then several studies have been conducted on the magnitude and scope of South-South migration. While large discrepancies exist between the various estimates, cf. Bakewell (2009), there seems to be agreement that South-South migration is characterised by high levels of irregularity; that most of the migration takes place across contiguous borders – and that hardly any migration exists between regions of the South (Lututala 2014); that both economic (from poorer to less poor countries) and political (from less stable to more stable situations) factors affect migration patterns; and that border control and movements registration in many countries in the Global South are characterised by weak capacities. Like other types of migration, South-South migration involves seasonal migration, transit migration, labour migration, family/network migration, circular migration, and forced migration.

Table 4.3 depicts estimates of world migration stock in 2007 and 2013, respectively. At first, it seems that South-South migration has decreased in importance relative to South-North migration from 2007 and 2013 (although absolute figures have increased), but aggregate figures are based on two different classifications of the Global South and registration of cross-border migration tends to be laxer in the Global South compared to the Global North. Hence, any comparison of aggregate figures over time is, at best, misleading.

Moreover, the phenomenon is mostly studied by approximation. In order to understand the processes at work it is therefore essential to disaggregate them and look at particular flows in more detail. Below, the Chinese migration to African countries will illuminate

Table 4.3   *Global migration stock 2007 & 2013*[7]

	MIGRANTS LIVING IN:						
	2007				*2013*		
MIGRANTS FROM:	South		North		South		North	
SOUTH	73.9 mn	44%	81.9 mn	48%	93.1 mn	38%	84.3 mn	34%
NORTH	4.2 mn	2%	29.1 mn	17%	14.2 mn	6%	55.7 mn	23%
TOTAL	78.0 mn	46%	90.9 mn	54%	107.3 mn	43%	140.0 mn	57%

Source: (Ratha and Shaw 2007; Ratha, Eigen-Zucchi, and Plaza 2016).

the politicised nature of this phenomenon and point to the historical developments that we have witnessed in the past century.

Interregional South-South migration: the case of Chinese migration to African countries

Chinese migration to African countries is often highly politicised – both domestically and internationally. As described in Textbox 2.1, numerous Chinese workers temporarily migrated to Tanzania and Zambia during the construction of the TAZARA railway. This was not the first time Chinese workers assisted in railway construction in Africa. According to Lututala (2014), under colonialism, 540 Chinese workers were recruited to build the Matadi-Kinshasa railway (1891–1898) and 600 Chinese manual workers were engaged in the construction of the Congo-Ocean railway (1921–1934). In South Africa, an estimated 3,000 Chinese were living and working in the Johannesburg area by the end of the nineteenth century. They operated large import firms, grocery stores, and laundries. At the beginning of the twentieth century, some 60,000 Chinese workers were 'imported' to work in the mines of Transvaal. However, three to four decades later, in 1940, only some 4,000 Chinese lived in South Africa, dealing mostly with commerce and trade (Pan 1999). In the years following Mao Zedong's rise to power in China he sent approximately 150,000 Chinese technicians to work on various aid projects (Park 2009). During the 1970s, several Chinese construction workers temporarily migrated to African countries to build palaces, stadia, roads, and railways. By far the majority of these workers returned to China, but a few stayed on and became the founding fathers of some of the construction companies that exist in African countries today. A distinctive feature of these companies is that they took up names such as 'African brothers' that masked their Chinese origin (Kragelund 2009b).

At the end of the twentieth century, a new wave of Chinese migrants began to arrive in select African countries. It was a heterogeneous group comprising medical doctors and agricultural personnel working on Chinese aid projects; construction workers and the sporadic miners working in state- and provincial-owned companies either seeking to internationalise or supporting Chinese aid projects; teachers; and a growing number of independent entrepreneurs either pulled to African

Plate 4.1 *4,200 Chinese mine workers, South Africa, early twentieth century*

countries to profit by providing for the growing Chinese communities that were emerging, or pushed out of China by the growing internal competition. This latter group consisted among others of craftsmen, traders, people working in the local restaurant business, people setting up traditional medical clinics, and sex workers (Haugen and Carling 2005; Ndjio 2009; Dobler 2008b).

Table 4.4 provides an overview of the magnitude of this new wave of Chinese migration to Africa. However, it is important to bear in mind that the figures are at best estimates and that different sources vary markedly. This being said, Table 4.4 on the one hand informs us that Chinese migration is not only a thing of the past couple of decades. In Mauritius and South Africa, numerous Taiwanese and Hong Kong Chinese worked in the textile and clothing industry in the 1960s – 1980s, cf. Hart (2002). On the other hand, it shows the growing ties between China and Africa really took off at the turn of the millennium. While Chinese migrants were hardly visible in Ethiopia by the year 2000, the number had grown to approximately 3–4,000 by the middle of the first decade of this millennium. Likewise, the number of Chinese people in Sudan numbered less than 50 in the year 2000.

Table 4.4 *Estimates of Chinese migrants in select African countries 1963–2007*

	1963	*2000 (CHINESE SOURCES)*	*2001 (OHIO UNIVERSITY DATABASE)*	*2003–2007 ESTIMATES*
ETHIOPIA	18	8	100	3,000–4,000
LIBERIA	27	45	120	600
MAURITIUS	23,266	40,000	35,000[8]	30,000
MOZAMBIQUE	1,735	500	700	1500
NIGERIA	2	10,000	2,000	100,000
SOUTH AFRICA	5,105	70,000	30,000	100,000–400,000
SUDAN	NA	45	45	5,000–10,000
ZIMBABWE	303	500	300	10,000

Source: Adapted from Mohan and Tan-Mullins (2009).

Some five years later, it had grown to 5–10,000. In Ethiopia, the influx is directly related to the Chinese investments in the leather industry (Gebre-Egziabher 2007) – in Sudan it relates to the massive Chinese investments in the oil industry (Patey 2014). Table 4.4 also depicts the poverty and paucity of data on Chinese migration.

Zambia exemplifies many of the processes at work in China-Africa migration: Chinese labourers first came to Zambia to construct infrastructural projects, ministries, and stadia and to work in friendship farms and factories. A few Chinese migrants stayed – either in Zambia or in the region – when China began to downsize its overseas activities in the late 1970s. They became central actors when Chinese migrants returned to Zambia in the late 1990s. The importance of Zambia as a migration destination is attributable to several concurrent processes. Zambia began a process of liberalisation and privatisation of state-owned enterprises in the 1990s. Most importantly, Zambia sold off its crown jewel, the Zambia Consolidated Copper Mine, to foreign mining firms including a state-owned Chinese mining corporation. Alongside the mine an explosives factory also came into Chinese hands. China meanwhile opened its first overseas bank in Zambia and slowly, but steadily, either began rejuvenating friendship farms, i.e. agricultural projects funded by the Chinese state in recipient countries, and factories or pushed new companies to internationalise – either via incentives or as part of development finance schemes. The result was that a large number of Chinese labourers were sent to Zambia to work on the state – or provincial-owned firms.

Chinese migrants bunched together in enclaves and began demanding local food stuff, entertainment, and services (Lee 2009). The result was an influx of artisans, people setting up restaurants, medical centres, casinos, and construction companies. Chinese migrants were no longer confined to the mines and the construction of feeder roads in the Zambian hinterlands; they became visible in the cities. In Lusaka, the capital of Zambia, numerous restaurants, lodges, and medical centres shot open. Most visible, however, were the Chinese small-scale traders in Kamwala market. A Chinese company had constructed the permanent structures of Kamwala and Chinese traders ran all the stalls in the central area. What was once only an occupation for Zambians had by 2005 been taken over by Chinese traders selling only Chinese-produced goods. Roughly simultaneously, an accident happened at the Chinese-owned explosives factory killing almost 50 people and the international media turned their focus on labour relations in the Chinese-owned mines. Taken together it led to a politicisation of the Chinese presence in Zambia: during the 2006 Presidential elections campaign Chinese migrants became scapegoats for all bad in the country. The main opposition candidate, the late Michael Sata, called the Chinese 'infestors' not investors and after the elections, won tightly by the incumbent, late Levy Mwanavasa, riots began in the centre of Lusaka leading among other things to a restructuring of Kamwala market using Zambians as fronts, under the pretext that Chinese no longer owned market stalls in Lusaka (Haglund 2009; Lee 2009; Kragelund 2009a, 2009b, 2012a; Hairong and Sautman 2012).

Since then, the number of Chinese migrants in Zambia has risen again. The actual figure is no less politicised now compared to 10 years ago:[9] estimates vary from 4–6,000 in 2007 (Park 2009), over 13–23,000 in 2012/13 based on careful analysis of immigration permits (Postel 2016) and 30,000 Chinese migrants working in the Lusaka markets alone in 2006 (Konings 2007) to 80,000 that Michael Sata claimed were in the country in 2006. A number of factors contribute to the large discrepancies. First, ever since the 2006 Presidential elections the figure has been highly politicised. Second, while entry permits into Zambia are carefully registered, this is not always the case with exit stamps, i.e. there is a real risk that immigrants either overstay their permits or never leave, meaning that the actual number of immigrants is higher than what is registered by the immigration authorities (Postel 2015). Finally, illegal border crossings from neighbouring countries are not registered.

Education: changing the hearts and minds of the next generation

For years, the Global North has formed the educational sector in the Global South: in many parts of Asia and Africa, the Global North created formal education, first with missionaries, then colonisers, and finally the donor community. In the post-World War II era the higher education system that was built in the Global South mirrored the system of the colonial powers. The focus was often on humanities and social sciences rather than technical (applied) competencies. In fact, many tertiary educational institutions in the Global South kept an affiliation to the former colonial powers well beyond formal decolonisation to 'guarantee' high academic levels (Selvaratnam 1988). In the past couple of decades, however, we have witnessed what Naidoo (2011) calls a critical disruption in the inter-hegemonic rivalry between 'traditional' and 'emerging' actors that may lead to major changes in the educational sector. Education now also features high on the agenda of all actors of the Global South (Chapman, Cummings, and Postiglione 2010) and except for a few OPEC donors support for educational development is a central element of all major Southern actors' aid programmes (Kragelund 2008).

Box 4.5 Internationalisation of Malaysian higher education through South-South Cooperation

Internationalisation, i.e. the integration of international aspects in the teaching, research, and service functions of an institution of higher education, is high on the agenda of most governments in the Global South (Adriansen, Madsen, and Jensen

2015). It seems to signal an efficient way of dealing with current socio-economic challenges, and refers to changing teaching standards, delivery of teaching services to institutions located in other countries, the use of ICT to offer programmes previously located in one country in another country, academic mobility, as well as collaborative research programmes.

In Malaysia, internationalisation of higher education is part of a transformation to a knowledge-driven economy. Via its National Higher Education Strategic Plan, the government of Malaysia aims to become an international hub for educational excellence. Unlike other efforts to become excellent, Malaysia is not only looking towards top-ranked universities in the Global North but also to universities in the Global South.

According to Wan and Sirat (2018), South-South internationalisation of higher education comes in a variety of forms in Malaysia. The Malaysian Technical Cooperation programme, for instance, dates back to 1980. In short, it is premised on the belief that social and economic development to a large extent relies on human resources. By definition, therefore, this programme goes beyond training in higher education but importantly it includes scholarship programmes to allow especially Islamic and/or neighbouring countries to benefit from the high quality of higher education in Malaysia. Of more recent origin is Malaysia's Global Reach programme that seeks to enhance higher education via exchange of researchers and alumni. Similarly, a public university in Malaysia has joined the Africa-Asia Development University Network. This network allows for joint supervision of doctoral students, publication of working paper series, and coordination of research initiatives of common interest to the regions.

Source: (Wan and Sirat 2018)

In particular, China and India put education high on the SSC agenda. Unlike the other vectors described above, educational collaboration directly affects the hearts and minds of the next generation of leaders in the Global South. This is especially the case for student scholarship programmes, as soft power theory '*presumes that students with pleasant first-hand experiences of life abroad will admire the host country's political system and, in turn, push politics at home in the direction desired by the country they studied in*' (Haugen, 2013: 318), but it also applies to other forms of education cooperation such as the establishment of educational units within the host society such as the Confucius institutes (see Textbox 4.6).

Educational cooperation has been a central part of India's aid programme since the mid-1960s when the Indian Technical and Economic Co-operation (ITEC) programme was established. ITEC seeks to make India's experience in agriculture and industry available to her partners in the Global South. This is done among other ways via training of technical personnel in India and study trips. Of late, India has furthered its SSC in higher education via, for instance, the funding of a two-year Gandhi-Luthuli Chair in Peace Studies at the University of KwaZulu-Natal, South Africa, and the inauguration of the Centre for Indian Studies in Africa at the University of the Witwatersrand, South Africa (Hofmeyr and Betty Govinden 2008).

China is equally keen to adopt this instrument in its collaboration with other countries in the Global South. China has used the FOCAC meetings (see chapter 1) to formalise this vector of its engagement with African societies. These meetings have been instrumental in forming partnership agreements between Chinese and African universities; developing science and technology partnerships; providing direct support to African post-doctoral students; training numerous Africans in China; and constructing technical educational facilities in Africa (King 2013). On top of this, China sends teachers to African secondary schools and universities to cater for language training and it establishes seminars and workshops for African high-level administrators (Harlan 2017), but most importantly it is fast developing Confucius centres and Confucius classrooms throughout the Global South (and North).

Box 4.6 The Confucius Institute in the University of Zambia

Confucius institutes are academic units located in a host country university governed and financed by the Office of Chinese Language Council International (often shortened to Hanban) in Beijing. In official parlance, they seek to intensify educational cooperation, support the development of Chinese language education, and increase mutual understanding. They are expected to increase the inflow of foreign students and researchers into China. Thereby, they aim to make Chinese a global language for teacher training, while simultaneously advancing trade and investment between the home and the host economy. To do this the Confucius institutes offer short- and long-term teaching programmes in Chinese, Chinese culture courses, film screenings, art exhibitions, and sports.

The first official Confucius Institute was established in Seoul in 2004. As of the end of 2013, a total of 833 Confucius collaborations existed worldwide – mostly in the Global North. Confucius institutes are built on twinning arrangements between two universities – one Chinese and one in the host country. In the Global South, most Confucius institutes are joint ventures between a Chinese university, Hanban, and a local university. In this format, Hanban provides start-up funding, books and teaching material and pays the salaries of one or two language instructors.

This was also the case in Zambia. The Confucius Institute at the University of Zambia was established in modest premises in July 2010 and teaching of the first class of 20 students began three months later. Since then, more than 2,000 students have attended classes at the institute. The institute has now moved to a very spacious two-storey building which will also accommodate staff from the central administration of the university financed entirely by Hanban. This allows for a large-scale expansion of the activities to cover numerous short-term courses and a degree programme in Chinese. Unlike other degree programmes at the university, however, the University of Zambia has neither a say over the pedagogy and curricula nor over the appointment of teachers. The Confucius Institute has changed funding arrangements as well as the mode of operation. Now it is a wholly Hanban operated and funded unit.

Hanban, of course, does not work as a charity organisation. The decision to build and fund a new building at the university came after the Ministry of Education in Zambia decided to offer Chinese at secondary level. Thereby, Mandarin and French have equal status in the Zambian educational system. Alongside the expansion, the Confucius Institute has also increased its efforts to be a local opinion former. This is done through news reports produced by Confucius Institute staff that feed into the growing presence of Chinese media in Zambia; through Confucius Institute shows on University of Zambia Radio; and through a formalised agreement with the Zambia National Broadcast Corporation to broadcast a television show entitled '*Get to Know China*'.

The establishment of the Confucius Institute at the University of Zambia should be analysed alongside other nations' widespread use of language and cultural encounters as a soft power tool in Africa. The promotion of French, German, and English language and culture has been an important task for institutions like the Alliance Française, Göethe Institute, and British Council since independence.

The Alliance Française is the main instrument through which French language policy is promoted in Africa. It functions under the French Embassy and operates throughout Africa. In Zambia, it has branches in five major cities. Through these branches, the Alliance Française offers French language lessons to the general public in Zambia and targets French teachers in Zambia, giving them the skills to promote knowledge about the French language throughout the country. Likewise, the Göethe Institute has just announced that it has initiated collaboration with the University of Zambia Language Centre to facilitate the spread of German as a foreign language in Zambia and Deutche Welle, Germany's international broadcaster, broadcasts throughout Africa communicating German points of view, such as liberal democratic values and respect for human rights. Finally, the British Council facilitates Zambian University students' studies in the United Kingdom and offers a range of school-linking programmes intended to further intercultural understanding and networking.

Source: (Kragelund and Hampwaye 2016)

Global governance: using economic power to change global power relations

The final vector of engagement is global governance. Just like the Bill and Melinda Gates Foundation has influenced the broader development arena via its focus on speed of delivery and cost effectiveness (Fejerskov 2015), SSC is also changing the UN system from within (Milhorance and Soule-Kohndou 2017). Equally importantly, key actors in the Global South actively seek to challenge the power structure within global governance. One of the most innovative challenges is the New Development Bank.

The New Development Bank was launched alongside a Contingent Reserve Arrangement at the BRICS summit in Brazil in June 2014. The bank took off with an authorised capital worth USD 50 billion intended to be lent to the five members of the bank as well as to non-members. As described in chapter three, one of the main aims of BRICS was to alter the power relations within the international financial institutions. In particular, BRICS aimed to alter the Special Drawing Rights, i.e. the quotas that countries are allowed to access when in financial need. Even if the BRICS had not been particularly successful in this endeavour, a change seemed to take place in 2012 when the IMF – following the global financial crisis – needed a bailout worth USD 430 million. In total, BRICS contributed USD 75 million towards the IMF. In return, 6 per cent of the quotas should shift from 'under-represented' nations to 'over-represented' nations. The US congress, however, postponed the approval of the reform and even though it was finally passed in 2015, the protraction of the process led BRICS to not only launch its own Development Bank but also to introduce its own Contingency Reserve Arrangement – a direct competitor to the Special Drawing Rights – that would enable BRICS countries to forestall short-term liquidity pressures (Gray and Gills 2016).[10] Moreover, BRICS members were dissatisfied with the extent to which the IFIs were able to meet the financing for infrastructure needs of the Global South.

Although the New Development Bank took a relatively long time to really take off, it is now perceived as an innovation in global financial

Plate 4.2 *Launch of the New Development Bank, Sandton, South Africa, 17 August 2017*

governance due mainly to the following four distinct features: rights and obligations are based on 'equality' among the five members (even if China can draw 25 times as much as South Africa); focus is on sustainable development via green infrastructure (in particular small-scale renewable energy projects); the capital base is built on bonds in BRICS national currencies; and there is an emphasis on speed in delivery (Cooper 2017).

Conclusion

This chapter has examined the main vectors of SSC engagement, i.e. development finance, trade, investments, migrations, education, and global governance, and shown how they are interdependent – often triggered by development finance. It has also revealed that even if we schematically are able to distinguish SSC vectors from vectors originating in the Global South, heterogeneity within the Global South (and Global North for that sake) is relatively large putting in question the dichotomy between South and North for each of these vectors.

In reality, what we see is that aspects of each of these SSC vectors resemble what the Global North is/was doing.

Discussion questions

- Describe how South-South development finance differs from North-South aid and discuss how South-South development finance affects other vectors of engagement.
- Explain how global trade is changing and how it affects the Global South.
- Consider why migration data are notoriously vague.
- Discuss how investments may further economic growth in the Global South.
- Discuss how and the extent to which the Confucius Institute resembles South-South Cooperation.

Web pages of interest

- AidData is a database of geocoded aid data. It provides data on a wide range of aid projects including the African Development Bank's projects in Africa, Chinese aid projects and Brazil's SSC: www.aiddata.org
- Financial Tracking Service provides geocoded data on humanitarian assistance – where they are from and where they are going: fts.unocha.org
- OECD's development finance statistics offer an overview of all the countries in the Global South reporting their development finance to the DAC and estimates of the magnitude of several Southern donors' development assistance programmes: www.oecd.org/dac/stats/non-dac-reporting.htm
- The Trade Law Centre, Tralac, is a South African-based organisation focusing on how regional trade integration, agricultural development, and changes in finance and development affect East and Southern Africa. It offers regular analyses of these issues and has published policy briefs and working papers informing the discussion: www.tralac.org
- UNCTAD stats provides data on trade in final products (see chapter 1), foreign direct investments, and economic trends: unctadstat.unctad.org/EN

Notes

1 Although India sometimes uses the term aid in formal documents it officially rejects the term.
2 One Southern donor, South Korea, was accepted as a full DAC member in 2010 thereby further blurring the distinction between the Global South and the Global North.
3 Gulrajani (2016: 7) reports that from 2008–2013, 60 per cent of ODA from DAC donors was disbursed via bilateral channels.
4 The Gulf states also comprise Bahrain, Iraq, and Oman.
5 Simply put, FDI can come in two forms: either as greenfield investments or brownfield investments. While the former denote a situation where the parent company constructs new facilities in the host economy, the latter denote a situation where the parent company purchases an existing facility in the host country.
6 The investment motives are not mutually exclusive; some firms may be both market and resource-seeking. Hence the total sum is above 100 per cent.
7 Figures are not comparable as 2007 figures follow World Bank classifications, i.e. South denotes low and middle income countries, North denotes high income countries (Ratha and Shaw 2007) whereas 2013 figures follow UN classifications, i.e. South here denotes countries classified as 'developing' countries and North denotes countries classified as 'developed' countries by the UN (Ratha, Eigen-Zucchi, and Plaza 2016). Mix of definitions seem to be the rule rather than the exception in studies of South-South migration. Bakewell (2009) compares UN and World Bank data (2007) and argues that by using UN data instead of World Bank data, South-South becomes 21 per cent lower, i.e. the comparable figure in Table 4.3 is 58.1 million South-South migrants.
8 1990 figure.
9 In 2015, a member of parliament in Zambia wanted to know the number of Chinese in the country. The official number was 13,000 (Chibuye and Mvula 2015).
10 It should be noted though that while the New Development Bank indeed is challenging established norms in global governance, the Contingent Reserve Arrangement according to Bond (2016) in fact empowers the IMF as countries in need of more than 30 per cent of its quota need to fulfil conditions stipulated by the IMF.

Further reading

Azmeh, S., and K. Nadvi. 2014. Asian firms and the restructuring of global value chains. *International Business Review* 23 (4):708–717. This article offers an analysis of a Southern-led global value chain. It offers insights into the effects of Asian firms' investments in Jordan and offers an analytical framework to understand these investments.

Bond, P. 2016. BRICS banking and the debate over sub-imperialism. *Third World Quarterly* 37 (4):611–629. In this article Bond offers a critical reading of both the BRICS Contingent Reserve Arrangement and the New Development Bank. In his reading, both resemble the Bretton Woods institutions already in existence.

Bräutigam, D. 2011. Aid 'with Chinese characteristics': Chinese foreign aid and development finance meet the OECD-DAC aid regime. *Journal of International Development* 23:752–764. Via an examination of two types of Chinese development finance, Bräutigam discusses the differences and similarities between Chinese and DAC development finance. She concludes that only a small part of Chinese development finance resembles ODA.

Gibbon, P., and S. Ponte. 2005. *Trading Down. Africa, Value Chains, and the Global Economy*. Philadelphia: Temple University Press. This well-researched and beautifully organised book offers an extraordinary analysis of how changes in the global economy – in particular the changing governance structure of Anglo-Saxon firms – affect African economies. By introducing the global value chains approach it presents a valuable tool to understand how economic globalisation affects development. Instead of scrutinising individual African countries, it examines trade in key agricultural crops.

King, K. 2013. *China's Aid and Soft Power in Africa*. Woodbridge: James Currey. In this well-written book King analyses a particular vector of Chinese aid engagement in Africa, namely aid to the educational sector. It offers an in-depth insight into how China over the past 60 years has used a variety of tools to change the hearts and minds of Africans.

Mohan, G., and M. Tan-Mullins. 2009. Chinese Migrants in Africa as New Agents of Development? An Analytical Framework. *European*

Journal of Development Research 21 (4):588–605. Despite the scarcity of reliable data, this article offers an initial assessment of the social, economic, and political effects of Chinese migration to African countries.

Oonk, G. 2013. Settled strangers: Asian business elites in east Africa (1800–2000). New Delhi: SAGE Publications India. This book is an important contribution to our understanding of South-South migration in historical persective. Oonk scrutinises the culture, economics, and politics of two centuries of South Asian migration to East Africa focusing in particular on the families, the so-called 'settled strangers' and how they have been embedded in different cultures simultaneously.

⑤ Actors of South-South Cooperation

The 'drivenness' of South-South Cooperation

Inspired by the Global Value Chain literature, Farooki and Kaplinsky (2012) introduced the concept of drivenness to understand how China and India affect the rest of the world. Unlike the Global Value Chain literature, however, drivenness is not referring to the governance structures that derive from sectoral characteristics, but rather points to the fact that the global tectonic shifts that we see in economic and political power affect other parts of the world. Stated differently, some countries/sectors/peoples are being driven intentionally or unintentionally by the global changes. Simultaneously, a variety of actors in the Global South actively seek to drive these very changes.

Governments in countries as diverse as South Africa, China, India, and United Arab Emirates are establishing aid agencies to coordinate and boost their aid programmes and they are setting up Export-Import banks to facilitate trade and investments with the rest of the Global South. These institutions are complemented by incentives schemes to facilitate the internationalisation of home country firms via foreign direct investments. Similarly, these same governments continuously expand their representation abroad: they build new embassies, sponsor trade fairs, and engage in grand bi- and multilateral summits. Thereby, they also facilitate the growing South-South migration. Moreover, they seek to change the hearts and minds of the next generation of Global South elites via education programmes, cultural exchanges, and expansion of national media. What is more, these countries also are the home of numerous state-owned enterprises that actively engage in trade and investments in other parts of the Global South.

Central governments are not alone in driving this development. Local governments especially in China also actively seek to further SSC.

This is done via investment packages, information campaigns, and 'friendship' arrangements.

Private actors are not at all puppets in the governments' overall strategy to boost SSC. In contrast, they are very active players in the current developments. Large companies both benefit from and drive the current economic growth in the Global South. They see new untapped markets, unexploited resources, and bid for tenders on projects financed largely by the IFIs. Smaller companies either operate as sub-contractors for their home-country transnational corporations in the host economy or are being pushed out of their home economy due to increasing competition, increasing wage levels, and increasing pressure to abide to environmental and social standards. Others are being pulled to countries in the Global South by new opportunities or generous incentive schemes.

Despite the growing interest in SCC, most accounts of the flows of money, goods, or people tend to treat them as just that: flows. Politicians, policy makers, and academics alike seem to be most interested in the magnitude of these flows (and whether they increase or decrease). They are less interested in who is performing the activities that lead to these flows. Equally important, most accounts of SSC see this as a one-way process – despite the official rhetoric of win-win, mutual benefit, and equality.

This chapter therefore digs deeper into these actors. It first distinguishes between government and private actors – and tries to unpack both groups. Then, by way of three examples, namely 'African traders in China', 'the leather industry in Ethiopia', and 'organisational learning from oil investments in Sudan', it opens up a more complex picture of drivenness and the actors involved than what is normally portrayed. Instead of engagements driven by one type of Southern actor alone, these examples demonstrate that as soon as a relation is set in motion, a variety of actors get involved, and that what seems to be one-directional often ends up being two-directional: goods, ideas, and people flow in many directions.

Government entities

The tendency of perceiving SSC only as flows driven by intangible actors is no doubt most pronounced in depictions of China-Africa relations. The lion's share of media accounts of China-Africa relations – and even some academic literature – has failed to disaggregate both 'China' and 'Africa'. Even if the views on whether this is good or bad – and for whom – differ radically, China is most often treated as a monolith. This is also the case for Southern commentators. China's state agencies, for instance, present China's presence on the African continent as the continuation of past engagement and as something unquestionably good – for both China and Africa. As demonstrated by Strauss (2013), this view is repeated by most private Chinese actors as well. The portrayal of the relationship by Western media used to be equally one-dimensional (Mawdsley 2008), although it is now possible to distinguish different perceptions (Strauss 2013). Likewise, most accounts of Africa neither distinguish between countries, nor between sectors, activities etc.

Most popularised versions of SSC leave out the complexity of relations. This is especially the case with regard to accounts of who is responsible for what. More often than not we hear stories about what 'China' is doing (or not doing) in 'Africa'; what 'Brazil' is up to in 'Lusophone Africa'; or what 'South Africa' is doing in the rest of the 'continent'. No one seems to bother to inform the reader who among the 1.3 billion Chinese people is taking the lead in the interaction with more than 800 million people living on the African continent. At best, it is the Chinese state/government doing this or the government led by former president Lula da Silva who initiated this or that initiative.

However, as alluded to above, neither is SSC driven by a single entity in each country nor is each country in the Global South a monolith. Rather, a plethora of central, regional, and local government entities are responsible for setting different aspects of SSC in motion. Sometimes these very entities also implement the policies/projects, sometimes they are implemented by a plethora of private, civil, and state actors. The complexity of this system is most easily illustrated with regard to the aid system. Unlike in many countries of the Global North, countries of the Global South have no single entity responsible for ODA- and OOF-like flows.

China, for instance, until 2018 had a Department of Foreign Aid, but this department neither had an autonomous role with regard to aid provision, nor was it the only institution dealing with development finance to the rest of the Global South (see Figure 5.1).[1] In short, the state council has the overall say in the Chinese aid system. It

Figure 5.1 China's aid and economic cooperation system (*until April 2018*)

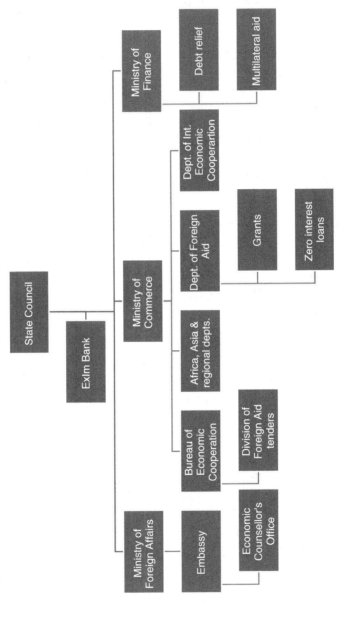

Source: adapted from (Bräutigam 2009: Figure. 4.1)

approves the yearly budgets and deals with politically sensitive issues. Directly under the state council are the ExIm Bank in charge of China's strategy to internationalise its businesses, the 'Going Global Strategy'; the Ministry of Foreign Affairs in charge of *inter alia* China's corps of foreign diplomats who advise the ministry on how much aid a particular country needs; the Ministry of Commerce in charge of several entities dealing with aid including the Department of Foreign Aid; and the Ministry of Finance in charge of debt relief and multilateral aid (Bräutigam 2009).

The Brazilian aid system is no less complex than the Chinese one (see Figure 5.2). Here, the Brazilian Cooperation Agency (ABC) is at the centre of most activities, but unlike aid institutions elsewhere, it hardly implements projects itself. Rather, implementation is outsourced to a variety of national and regional entities. One of the most important actors in this regard is the Brazilian Agricultural Research Cooperation (EMBRAPA) in charge of transfer of technology in the agricultural sector, the Ministry of Health in charge of public health related projects, and the Ministry of Social Development and Fight Against Hunger in charge of implementing Brazil's 'Zero hunger' programme across the Global South (Faria and Paradis 2013).

SSC involves several vectors of engagement beyond aid – even if development finance often facilitates these vectors. While the governance structure of these vectors is not as complex as the aid system described above, one single government unit alone seldom orchestrates the engagement. Large-scale Chinese state-owned enterprises, for instance, are not autonomous even if they manage their day-to-day operations without involvement. In the words of Haglund (2009: 632): '*The Chinese government does not get involved in the day-to-day operations of overseas Chinese enterprises but has a strong interest in setting policy objectives and monitoring firms' alignment with them.*' This is typically done via party committees within the enterprises; policy suggestions to guide company behaviour; and threats of sanctions if they misbehave. Moreover, these companies are involved with the ExIm Bank to finance their overseas operations, with the Economic Counsellor's Office to manage day-to day challenges in the host economy, and with the Ministry of Commerce. Moreover, it is important to further disaggregate the state-owned enterprise category into public institutions, central government-owned, and local/ provincial government-owned enterprises.

Figure 5.2 *Brazilian Technical Cooperation*

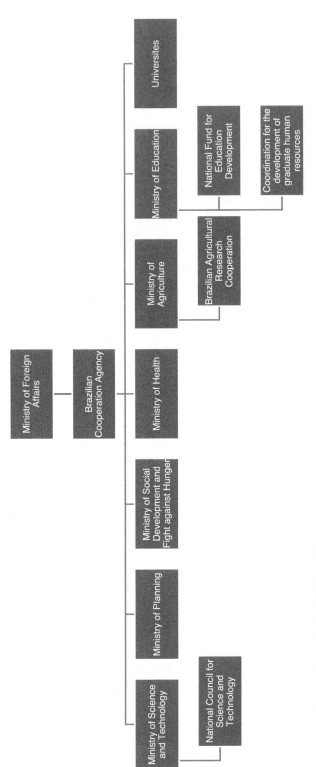

Source: adapted from (Faria and Paradis 2013: Figure 2).

Box 5.1 Female Indian peacekeeping forces in Liberia

Most of the peacekeeping forces deployed by the UN in the Global South originate in the Global South.[2] Pakistan, Bangladesh, and India are among the most important contributors to UN peace support operations. This comes as no surprise as these countries are characterised by a surplus of 'boots on the ground' military supply while simultaneously both the states that deploy the soldiers and the soldiers themselves can benefit from the deployment. However, an all-female South-South peacekeeping unit is original.

For more than half a decade participation in peacekeeping operations has been among India's most important contribution to the UN. India earns international recognition by this significant contribution (approximately 10 per cent of the total number of 'blue helmets') and since the 1970s, India has experimented with all-female formed police units and all-women police stations in India. Hence, when the UN as a consequence of the realisation that sexual harassment and sexual violence often is an integral part of conflict situations (and unfortunately also of many peacekeeping operations) began discussing gendered approaches to peacekeeping and the formation of an all-female peacekeeping unit, India was the obvious country to turn to.

In August 2003 a peace agreement was signed in Accra, Ghana, that ended almost one and a half decades of civil war in Liberia (interrupted by a couple of years of peace in the late 1990s). Not only had this period been extremely brutal, the country had also experienced an epidemic of sexual violence against women. Liberia was thus an ideal case to test the effects of all-female peacekeeping units: female soldiers were perceived as less threatening; and the presence of female soldiers was seen to empower Liberian women.

The 125 – person all-female unit is made up of members of India's Central Reserve Police Force. It was deployed in Monrovia, Liberia, in February 2007. At first, the unit was only deployed for six months, but tenure was soon extended to a full year after which a new group of female soldiers took over. Since the first group arrived in Monrovia in February 2007, ten new contingents have been sent from India to Liberia.

Their duties include patrolling the capital, Monrovia; keeping guard at the Ministry of Foreign Affairs; providing on-the-job training to the newly formed Liberian National Police; and of late, they have also taken charge of former president Ellen Jonson Sirleaf's security. No doubt, the all-female unit was perceived as a success in terms of being less threatening, but it is unclear whether it managed to empower Liberian women. In fact, sporadic evidence seems to suggest that the Indian soldiers had little contact with the local population. Moreover, cultural and geopolitical differences between India and Liberia seem to augment the distance between the two groups.

Sources: (Henry 2012; Pruitt 2016)

Private entities

The private actors in SSC cover an even more diverse group than the government entities described above. No doubt private firms in all shapes and sizes make up the largest group of SSC actors. Development finance from as diverse actors as India, UAE, and Brazil use private firms to implement some of their policies; and private firms from all over the Global South win tenders for construction, maintenance, and repair of infrastructure projects funded by the Global North and South alike. Trade is to a large extent carried out by private firms and individual traders (see also Textbox 5.4). The same goes for migration. Only education and some investments are carried out by state actors.

Box 5.2 South-South Volunteerism

The role of volunteerism in international development has received a lot of attention (Devereux 2008). The lion's share of this work, however, focuses on volunteers from the Global North working in the Global South. While this is important and the scale and magnitude is increasing year by year, it misses the important contribution of South-South voluntarism. South-South volunteerism involves among others faith-based volunteers, volunteers from international voluntary networks, and corporate volunteers.

By far the most important type of South-South volunteerism is the type brokered by international organisations. Principal among these is the United Nations Volunteer Programme (UNV) that seeks to promote peace and development worldwide via volunteerism. From its inception in 1971, it has aimed to promote South-South Collaboration – first of all by physical dispatchment of volunteers from one country to another and, since the turn of the century, also via online volunteerism. Currently, more than 80 per cent of the volunteers at UNV are from the Global South (UN Volunteers 2018). These international organisations also include the International Federation of the Red Cross, Voluntary Service Overseas, and Volunteers in Conflicts and Emergencies Initiative.

According to Baillie Smith, Laurie, and Griffiths (2018) this type of South-South volunteerism is simultaneously characterised by an imaginary of 'sameness', i.e. volunteers from the Global South perceive that their experiences and identities are relatively easily transferred to other parts of the Global South due to a shared identity as originating in the Global South, and a recognition of increasing heterogeneity within the Global South. This heterogeneity leads to new hierarchies as volunteers from the Emerging South work and operate in the rest of the Global South.

On a somewhat smaller, but growing, scale is South-South corporate volunteering. Thus far, only a few companies from the Global South have engaged actively in this, but Tata Group, India, may be the exception to this rule. The consulting arm of the business group, Tata Consulting Services, has established a particular structure as a '*vehicle for engaging employees and their families in activities both within the company and in the community, providing an "out of cubicle experience" for personal as well as professional development*' (Allen and Galiano 2017: 106). It is perceived as an important component of the company's genetic material, the so-called 'Tataness'. Likewise, the Latin American financial sector is rapidly developing its corporate volunteering programmes. In some countries, like Brazil and Colombia, they even established corporate volunteer councils to facilitate sharing of experiences.

Sources: (UN Volunteers 2018; Devereux 2008; Allen and Galiano 2017; Baillie Smith, Laurie, and Griffiths 2018).

It comes as no surprise that the majority of Brazilian, South African, and Indian firms investing in the rest of the Global South are privately owned. What is more surprising – at least for those who have only followed the debate in the media – is that the majority of Chinese firms investing in the rest of the Global South are also private. This is a relatively new phenomenon. According to Gu (2009), the growth of Chinese private enterprises only really took off in 2005. Since then, however, the number has increased dramatically. In Africa, this growth seems to follow a three-pronged process: first, private entrepreneurs trade and if successful, they begin to invest. Depending on the infrastructural developments and the technological capabilities in the host economies, the new investors bring in other firms to construct infrastructure and supply essential inputs and services. The third phase according to Gu (2009) is the agglomeration phase where investors concentrate in a spatially delimited area to benefit from the infrastructural investments and the closeness to suppliers.

According to Xiaofang (2015: figure 2), a major difference between the investments from Chinese state-owned firms and private firms is their sectoral focus. While state-owned firms primarily target the extractive industry and construction, private firms primarily target investments in manufacturing and trade and logistics. This sectoral difference correlates both with the degree of capital-intensity and the strategic importance of the investments, i.e. the more capital intensive and the bigger the strategic importance the higher the chance that the investment is made by state actors.

Box 5.3 The limited role of civil society organisations in SSC

While civil society organisations have played a tremendous role in alleviating poverty, offering basic education, providing food and security, and dealing with youth unemployment in many 'emerging' economies they have yet to play this role in these economies' activities in the rest of the Global South.

In India, for instance, the government led by Prime Minister Narendra Modi officially urges both the private sector and civil society to take part in India's external activities and the Ministry of Foreign Affairs in New Delhi has formed an agency, the Development Partnership Administration, to promote and coordinate all India's SSC activities. Moreover, the government in January 2013 formed the Forum for Indian Development Cooperation also under the Ministry of Foreign Affairs to bring together experiences from India that could be used elsewhere. However, a combination of legal impediments, limited external experience, and huge internal challenges related to poverty and inequality coupled with government hostility in some cases means that Indian civil society in reality is yet to become a player in SSC.

Like India, South Africa is also characterised by a vibrant civil society especially with regard to social justice struggles and democratisation but unlike in India, it has also been active outside its own borders. However, despite the fact that some of these activities have been funded by the government-run African Renaissance Fund, South Africa's civil society is still not an important player in South Africa's SSC activities.

This is also the case in Brazil where historically civil society has played an important internal role in transforming the Brazilian society and an active external role against global injustice. While the Brazilian government occasionally has invited the domestic civil society to influence its SSC policies, the civil society more broadly is yet to play an active role in implementing its policies and programmes abroad. Exceptions do exist, though, like Conectas, a Brazilian-based NGO that seeks to advance human rights across the Global South. Likewise, think tanks have increased their influence lately and service-NGOs also seem to be included in SSC funded by the Brazilian government

So far, therefore, the capacity of the domestic civil society – developed via ongoing struggles for social justice at home and abroad as well as via collaboration with Northern development organisations – has not been put in systematic use in SSC. This, however, is not surprising as the room for manoeuvre for rights-based civil society organisations is currently closing down in the South as well as in the North.

Source: (Pomeroy et al. 2016)

Two-way interaction

Accounts of SSC development in general do not differ much from most accounts of North-South development: they are about how the rich/powerful big country/government firm/organisation engages with the less affluent/powerful state/institution or how the latter is affected by the activities of the former. Only seldom is the interaction in focus and even less often is it acknowledged that relations change over time and new actors take over or enter into the relationship.

The following three subsections will provide insights into the multifaceted nature of these interactions. The first section takes us to the east coast of China where African traders take the opportunity to benefit from globalisation by ordering manufactured goods directly at the gates of the factories and warehouses and providing feedback to the producers on the tastes in their country of origin. The second section moves from the individuals to the firms and tells the story of how Ethiopian leather and shoe manufacturers have learned from Chinese companies and thereby been able to reverse a tendency of import of these goods to export into China. Finally, the third section tells the story of how Chinese and Indian oilmen have learned from their engagements in Sudan and how this knowledge is used at home in China and India, respectively.

African traders in China: 'Chocolate city' in Guangzhou

Trade of manufactured consumer goods, i.e. garments, textiles, consumer electronics, and household goods, between China and African countries take three distinct forms, namely wholesale and retail trade in African cities led by Chinese entrepreneurs; wholesale and retail distribution in major African ports directed by Chinese traders whereupon African traders take over and distribute the goods; and wholesale export from China led by African traders (Lyons, Brown, and Zhigang 2012).

While the focus in academic literature and the media primarily has been on the two former forms – and the Chinese involvement in this, there is now growing recognition that Africans also go to China, i.e. the third form. While Chinese traders are spread throughout the African continent, African traders seem to concentrate in three cities, namely Guangzhou, Foshan, and Yiwu in the South Eastern part of China. Due to the high concentration of African traders in Guangzhou, in particular, it is sometimes referred to as 'Chocolate City' or 'Little Africa'. In fact, African traders concentrate in two distinct areas of the town, namely Xiaobei and Sanyuanli – both close to the railway station. 'Chocolate City' in reality only refers to Sanyuanli, whereas 'Little Africa' refers to both areas. Trading takes place either in select warehouses or in districts that specialise in particular commodities, e.g. one district for electronics, another for toys, and a third for shoes and socks (Bork-Hüffer et al. 2016; Lyons, Brown, and Zhigang 2012; Cisse 2015).

The African traders in this part of China are a very heterogeneous group. They include the 'visitor' who comes twice a year for one or two weeks to buy textiles, leather goods, furniture, and consumer electronics for his/her relatives, friends and colleagues with the aim of making an additional income (see also Textbox 5.4); the 'trader'/'semi-settled', who comes as often as possible and stays for longer periods of time to work with producers to cater for the special demands at home, and with customs officers to make sure that everything is shipped as planned; and the 'entrepreneur' who sets up wholesale shops/factories

Plate 5.1 *African traders in Guangzhou, China*

in and around Guangzhou to sell to the African 'visitors' and 'traders'. As foreigners in China are not allowed to own majority parts of businesses, they link up with Chinese counterparts – either through joint ventures or through marriage.

Estimates on the number vary greatly – from some 1,500 African traders living and/or visiting in Guangzhou per year to close to 100,000 traders per year (Castillo 2016; Bodomo and Pajancic 2015). A number of factors account for the huge differences between these estimates, including the politicised nature of migration in China; the lack of a central immigration unit in China – both at provincial and state level; the fact that many visitors overstay their visas; the unclear definition of 'Africans in Guangzhou/China' (does it include people of African origin or only people coming directly from an African country?; does it refer to the African continent or only sub-Saharan Africa; does it refer to mainland China or China and Taiwan?; does it include 'Africans' living in Foshan but operating from Guangzhou?; does it include all travellers or only the ones that either stay for a certain amount of time and/or come back on a regular basis?); and the mix of motives that bring Africans to China, including students, professionals, tourists, and traders, cf. (Bodomo and Pajancic 2015).

There does seem to be agreement, though, that there is a gender and age bias among the African traders/entrepreneurs in China. By far the majority of Africans staying in Guangzhou, Foshan, Yiwu, and other megacities in China are relatively highly educated young men. Moreover, traders seem to predominantly originate in West Africa (Nigeria, Ghana, Mali, Guinea, and Senegal) and East Africa (Tanzania, Kenya, and Uganda). Traders originating in South Africa, Zambia, and Angola are much scarcer.

Despite the lack of precise data on the magnitude and scope of African trading activity in China, the literature on the subjects informs us about some of the processes that this activity sets in motion. First, there seems to be no doubt that the centres of production and trade on the east coast of China benefit from the increased economic activity that the African traders create. Not surprisingly, therefore, the city council of Yiwu, for instance, has speeded up the process of establishing a customs office, an international logistics centre, and an international banking system – all to enable the smooth operation of the production and wholesale business in town (Cisse 2015). This, however, does not ensure that being a trader/entrepreneur from the African continent on the east coast of China is straightforward. In contrast, the literature is rich on cases of surveillance and control imposed on the Africans by the Chinese state. Second, therefore, we also have a relatively clear picture that most city councils do not welcome the large inflow of foreign traders and more often than not they are the focus of large-scale policy raids to curb illegal migration. In 2009, for instance, such a policy raid – in which two African traders were seriously injured – led to large-scale protests in the African communities in Guangzhou (Lyons, Brown, and Zhigang 2012; Castillo 2016). Finally, we also know that despite the sometimes-hostile environment, African traders return because they make a profit from it (see also Textbox 5.4).

Box 5.4 Making ends meet by trading with Chinese goods in Lusaka

Not all the trade on the east coast of China is conducted by professional traders/entrepreneurs. In fact, a large share of the goods that enter African economies from other parts of the Global South are traded by civil servants or people with full time employment in the private or civil sector who seek to make ends meet by travelling to cities like Guangzhou or Yiwu to buy goods they resell to friends and relatives back home.

Hope lives with her two children in a two-bedroom apartment in one of the middle-class residential areas in Lusaka, Zambia. She now works in an international health NGO with offices throughout the country, but previously she worked in a Chinese organisation in Lusaka. During the past five to six years she has several times benefitted from the short-term training programmes offered for instance by the Indian and Chinese governments to employees of the Global South.

The combination of this international exposure and the recognition that several of her current colleagues added to their income by trading goods from China made her decide to pursue this income-generating activity herself. She began a process of mapping the demands and tastes among her friends. It soon turned out that what her friends really strived for were clothes, accessories, and consumer electronics. She took a week off from work, bought a return ticket to Guangzhou via Dubai, and linked up with a Chinese taxi driver cum trade agent who had been recommended by a colleague who had long experience of trading in China.

During her first trip to China, she spent roughly half of her time purchasing women's leather jackets, handbags, hair extensions, and mobile phones from the shops in and around the Tian Xiu Building in downtown Guangzhou – all of which were brought back to Zambia in her suitcases. The other half of her time, her taxi driver cum trade agent took her to the wholesalers and factories outside the central part of town, which produce and sell white goods, furniture, carpets, and the like.

Once safely returned to Lusaka, Hope began selling her goods. Based on her initial screening of demands and tastes she contacted relatives, close friends, and colleagues to persuade them to buy the stuff she had procured. Within a week, most of the goods that she had brought back from China had been sold. The rest she offered to acquaintances via Facebook. Soon after everything was sold-out and her profit margin was high: not only was she able to finance the air ticket, the hotel, and other travel expenses, she also had enough money for a new ticket to China and even some money to invest in new goods.

Her success combined with her newly gained knowledge about domestic appliances and furniture products for sale in the outskirts of Guangzhou made her decide on another strategy for her next trade trip. This time she approached potential buyers of sofas, white goods, and electronics and told them that she planned another trip and that, based on a 30 per cent down payment, would be willing to buy specific goods for them. She had photos of the leather sofas, carpets, ovens etc. for sale in the wholesale shops that she had visited in Guangzhou. Soon she had enough orders to fill a small container and off she went. Again, she filled her suitcase with textiles, clothes, and accessories. But for some hassle with customs officers in Zambia, it was once again a successful and highly profitable trip – indeed so profitable that she could buy a second-hand Japanese SUV instead of her two decades old Toyota Corolla. All her friends that had requested white goods and furniture paid for their orders and as with her first trip, Hope quickly managed to sell what she had brought with her in the suitcase to Facebook acquaintances.

Due to busyness at her workplace, Hope postponed her third trip to China to the beginning of 2017. She copied her strategy from the second trip, i.e. pre-orders of expensive goods based on partial down payment. Unlike the two first trips, however, the third was in no way a success: prices had risen in China, competition had increased in Lusaka, and the Zambian economy was experiencing a deep recession leading to a devaluation of the local currency, the Kwacha. In effect, goods were more expensive than anticipated in Guangzhou; more people offered similar goods in Lusaka; and most importantly, her customers did not have enough money to pay for the goods they had ordered. She ended up selling some goods at (too) discounted rates to close friends and trying to sell the rest via garage sales marketed via Facebook. The result, several months later, was a loss. Almost all imports were sold but profits were low or negative.

Source: Personal interview with Hope, Lusaka, 9 August 2017.

The shoe and leather industry in Ethiopia: creative destruction by Chinese investments?

Most accounts of the effects of SSC forget to consider the fact that what at first seems destructive may later turn developmental. An exemplary case of how time (and deliberate interventions) may change our assessment of effects of SSC is the shoe and leather industry in Ethiopia.

Until the beginning of the 1990s, high tariff and non-tariff barriers protected this industry from international competition. This meant that thousands of small- and medium-scaled Ethiopian shoe producers existed by the turn of the century. Besides protection, they benefitted from the high number of livestock and the good quality of highland sheepskins (Bräutigam, Weis, and Tang Forthcoming). When tariffs on imported shoes were reduced to 35 per cent in 2001 the situation for the local shoe and leather industry changed overnight. The increase in the total value of leather footwear imports illustrates the dramatic change that producers experienced. In 2000, the total value of leather footwear imports in Ethiopia was USD 15,595. In 2001, the value was USD 1,953,102. To make matters worse, approximately 90 per cent of these imports originated in China (Gebre-Egziabher 2007: Table 4). Exports of footwear to China – or any other commodities for that matter – at this point in time were non-existent (Gebre-Egziabher 2009).

Chinese footwear was cheaper, of better design, and perceived to be of better quality than locally produced footwear. Therefore, it came as no surprise that the Ethiopian footwear sector was hit hard by the liberalisation of trade. In particular, micro- and small enterprises were affected negatively. Firms either downsized radically or closed all together. Medium- and large-scale companies also experienced a blow. For example, the largest manufacturer at the time, Kangaroo, experienced a 50 per cent loss in sales revenues as a consequence of the imports from China (Bräutigam, Weis, and Tang Forthcoming: 5).

However, it was not long before some firms were able to benefit from the situation. The turnaround is visible among small firms serving the domestic market as well as larger firms exporting footwear. Probably the best example of the former is the open-air Merkato footwear cluster in Addis Ababa. Prior to the reduction of tariffs approximately 500 firms existed in the cluster. In 2001, this figure went down but already by 2005, the number had reached 1000 and by 2008 the figure was approximately 1500 (Gebreeyesus and Mohnen 2013: 305). The revival of the micro-firms was largely a result of imitation of the Chinese firms and innovation of products and processes (quality, design, variety of products, and investments in machinery).

The micro-firms, however, were not alone in the sector. Also in the footwear value chain, but spatially separated from Markato market, were the medium-to-large Ethiopian firms producing tanned leather, leather products, and leather shoes. They faced other, deeper, problems than the ones in Markato, including poor quality of hides, weak institutions, and inefficient marketing. Like the micro-firms they had been hit by what Bräutigam, Weis, and Tang (Forthcoming) call the 'China chock'. The Ethiopian government soon realised that they could not 'just' imitate their way back to success. Therefore, it crafted Ethiopia's Industrial

Plate 5.2   A shoe shop in the Markato market, Addis Ababa, July 2014

Development Strategy in 2002 that focused on the links in the entire chain from livestock via skins and hides to leather and footwear.

The idea was to benefit from attracting advanced foreign firms to Ethiopia and thereby, benefit from technology and knowledge spill-overs, from access to foreign markets, and access to capital. First, Ethiopia turned to its traditional markets in Europe, but even if agreements were made with both Italian and German firms, the arrangements did not work out. Instead, President Meles Zenawi went to China to persuade Chinese firms to set up production in Ethiopia. This visit led to a number of Chinese producers settling in and around Addis Ababa and kicked off the process of upgrading by learning and collaborating among the Ethiopian medium- and large-scale footwear producers. To speed up the process even more, the government of Ethiopia in 2008 introduced a number of export taxes that made it relatively more expensive to export unprocessed skins and relatively more profitable to export manufactured footwear and leather goods. Essentially, this forced the tanneries to import capital equipment and expertise to facilitate a transformation from producing (and exporting) unprocessed leather to producing processed consumer goods (Bräutigam, Weis, and Tang Forthcoming).

Learning from investing in Africa: Indian and Chinese oil companies in Sudan

The literature on the geopolitics of Chinese and Indian oil investments in Africa is vast (De Oliveira 2008; Klare and Volman 2006; Taylor 2006; Mawdsley and McCann 2011; Carmody 2011; De Oliveira 2015). This body of literature has paid particular attention to the similarities and differences between oil investments from the Global South and oil investments from the Global North; on what has driven the investments; and the effects on oil prices and the consequential economic and political impacts on the oil-exporting countries of the Global South. Much less, however, is known about how experiences from investing in countries characterised by civil wars, political instability, and large-scale corruption have affected the oil companies from the Global South and how they have used this experience to build capabilities that could be used to profit from investments elsewhere.

While oil companies originating in China and India were almost absent from the global scene at the beginning of the 1990s they are now global players competing head-to-head with oil giants from the Global North (and with each other) for new contracts. This development stems partly from technological leap-frogging and political support and economic subsidies from their home governments. But it also derives from the organisational changes that these companies had to implement in order to cope with the rapidly deteriorating political situation in Sudan and South Sudan that ultimately made them internationally competitive (Patey 2017, 2014; Large and Patey 2011). In the words of Patey (2017: 757): '*Sudan and South Sudan represented an international training ground that shaped their competitiveness and global strategies.*'

When Chinese and Indian oil companies entered the world scene they were weak financially, technologically, and organisationally: they were overstaffed; they lacked international experience, their foreign language skills were almost non-existent (for Chinese companies in particular), and they had no risk-management strategies. These weaknesses were scaled up by the fact that they at first were forced to invest in high risk, low profit places where oil giants from the Global North had no interest or were forced to divest due to political uproar at home. Sudan was one such place (Patey 2009).

Both for Chinese and Indian oil companies Sudan became an important learning spot. Thousands of Chinese and Indian workers were sent to Sudan to negotiate the oil deals, construct the facilities, and run the oil service companies and subsidiaries supplying essential goods and services to the oil companies. These workers not only included tradesmen but also white-collar workers who later took up senior management positions in the headquarters in Beijing and New Delhi and thus transferred the knowledge they had acquired to key decision-makers in the headquarters. Moreover, both Chinese and Indian oil companies built much needed organisational capabilities through operating companies both up- and downstream from the original oil field investments.

Chinese and Indian national oil companies received great economic and political support from their home governments (though the exact scope and magnitude differed in China and India) who perceived oil investments in Sudan both as a way to secure much needed oil to further economic development at home and as a way to continue the ongoing rivalry between these two Asian giants. The National Congress Party in Sudan was not a bystander in this competition for oil. According to Patey (2014: 147f), it played a very active role in playing out the two giants against each other and thereby increasing its own room for manoeuvre.

However, neither the Chinese nor the Indian government were able to protect their oil companies from the political instability they came to experience as a consequence of the Sudanese government's violent counterinsurgency in the Darfur region of Sudan. For a long time, this was no problem for the Asian oil giants as the Sudan armed forces were willing and able to protect their oil investments, but soon the Sudan armed forces were occupied elsewhere and various opposition groups to the National Congress Party began fighting the government outside Darfur – among other places – on its joint ventures with Chinese and Indian oil companies. This led to the development of a variety of company security measures, including internal training exercises; hiring private security firms; and collaborating with local civil society to fend off future attacks.

The story of Asian oil investments, therefore, is not only about geopolitics, neo-colonialism, resource curse, and corruption; it is also about how investments in a conflict-ridden part of the world changed the organisation of the national oil companies in China and India and thereby enabled them to compete (and collaborate) with Northern oil giants in places characterised by fiercer competition and easier access to oil. More than a decade of operating in an increasingly hostile environment taught these companies new strategies that were first transferred back to headquarters and then applied in operations all over the world.

Conclusion

This chapter has presented some of the main actors in SSC. It has shown that even if states play an important role in SSC they are not the only actors of importance. Rather, the lion's share of SSC activities are performed by a plethora of private actors including large-scale multinationals originating in emerging economies like Brazil, China, and India; large companies practising internationalisation strategies in countries characterised by low levels of competition either because laws forbid (some) Northern actors to invest or because these host economies are perceived to be too risky/unprofitable to invest in; small- and medium-sized enterprises pushed out of their home economies due to increased competition or pulled to a new host economy due to a combination of business possibilities related to state interventions (new roads, new hydroelectric dams, or new harbours) and opportunities created by the multinational corporations from their home economy; and individual traders, craftsmen, and farmers who are attracted by new business opportunities.

Private sector actors engage with – and benefit from – a multitude of state actors both at home and in their new host economies. This chapter has conveyed how the state in the Global South is everything but a monolith. In contrast, it consists of a multitude of entities that sometimes collaborate – and sometimes compete for funds and recognition.

Finally, and maybe most importantly, this chapter has argued that most SSC is neither one-directional nor static. Rather, SSC is an arena of connections set off by a development intervention, a tender, or an investment and which then attracts numerous actors that in one way or another benefit from the exchange. Also, this arena is ever changing and what may at first seem disadvantageous for one group of actors may over time turn out to be beneficial.

Discussion questions

- Discuss how government entities in the Global South relate to private and civic actors.
- Consider why it is important to follow South-South engagement over a longer period of time if we are to understand the effects of this engagement.
- Discuss why the civil society of the Global South hitherto has played a very limited role in SSC.
- Explain why the organisational setup of state institutions matter for our understanding of SSC.

Notes

1 In April 2018, the head of a new Chinese aid agency, the State International Development Cooperation Agency (SIDCA), was appointed. Although the main aim of SIDCA is clear, namely to lay out strategic guidelines and policies on China's foreign aid,

the structure of SIDCA as well as its precise coordinating role vis-à-vis the line ministries (most importantly Ministry of Commerce, Ministry of Agriculture, Ministry of Finance, and Ministry of Foreign Affairs) is still unclear. Thus far, aid related announcements still come from the Ministry of Commerce but the establishment of SIDCA suggests that China in the future will boost aid's role as a foreign policy tool.

2 UN Peacekeeping is by definition global and hence not solely SSC. It is included as an example of SSC here as most conflicts happen in the Global South and most of the forces deployed originate in the Global South.

Further reading

De Oliveira, R. S. 2015. *Magnificent and beggar land: Angola since the civil war*. Oxford: Oxford University Press. A detailed and lucid account of the political economy of postwar Angola. It offers a well-informed analysis of how oil rents affect development and how China's demand for oil has furthered this process.

Gu, J. 2009. China's Private Enterprises in Africa and the Implications for African Development. *European Journal of Development Research* 21 (4):570–587. This article takes the reader's attention away from China's state-owned enterprises in Africa. Instead it offers an insight into the myriad of private enterprises from China that currently invest in African economies.

Patey, L. A. 2014. *The New Kings of Crude: China, India, and the Global Struggle for Oil in Sudan and South Sudan*. London: Hurst. In this highly entertaining and well-researched book Patey takes the reader to Sudan and South Sudan through the lens of Chinese and Indian oilmen. He demonstrates how they use the experiences from operating in conflict-ridden countries to change governance structures in the oil majors at home.

6 Effects of South-South Cooperation on development 'as it was'

The North's initial reaction to SSC: from fear to cooperation and attempted capture

One of the first official recognitions of the emerging changes in the international development cooperation arena came from the then chair of the DAC, Richard Manning, in a speech that he gave to an 'All-Party Parliamentary Group on Overseas Development' in the spring of 2006. Here, Manning on the one hand pointed out that the emergence of Southern development actors would increase the availability of funds for development. On the other hand, he raised concerns about the following threats to development: increasing debt burdens for Southern lenders; postponement of (liberal) political and economic reforms due to the absence of conditionalities; and the waste of scarce resources thanks to unproductive investments (Manning 2006).

Whether this indeed was the correct reading of the situation has been subject to much debate (see e.g. Woods 2008; Paulo and Reisen 2010). Notwithstanding this controversy, the speech – and the realisation that the DAC was no longer the development aid hegemon – sparked off the development of a number of institutional arrangements and platforms that all sought to encourage Southern development actors to assimilate into the 'established' aid system. These arrangements included *inter alia* the China – DAC Study Group, the DAC Working Party on Aid Effectiveness (WPAE), and a multitude of trilateral development cooperation (TDC) arrangements[1] (Kragelund 2015). Furthermore, the 'established' donors of the Global North were keen to keep the focus of aid on poverty and social and economic development in recipient countries – not mutual benefit; to keep the link between economic and political reforms and aid – the use of *ex ante* and *ex post* conditionalities; to keep aid untied from purchases of products and services in donor countries; and not least, to keep their own legitimacy in times of growing aid fatigue. They therefore had to come up with ways of cooperating to maintain their relevance (Zimmerman and Smith 2011).

The China–DAC Study Group is primarily a way to cooperate. Essentially, it aims to facilitate mutual learning between China and the DAC. This is done via round tables that focus on aid quality and aid management and via policy symposia on issues such as agricultural development, infrastructure development, private sector participation in development, and how to monitor and evaluate aid interventions. The idea is to build trust and share experiences and thereby provide a basis for closer collaboration.

In contrast, the WPAE that took off in 2003 as a means to fulfil the eighth Millennium Development Goal (forming a global partnership for development) was primarily a means to assimilate the 'new' donors into the established system. The WPAE was organised around a series of High-Level Fora (HLF) on aid effectiveness beginning in 2003 in Rome, Italy, and ending in Busan, South Korea in 2011 (see also below). Originally, it was only a forum for Northern donors but soon it came to include Southern recipients as well, and with the realisation that the DAC had lost its absolute power to define the future of development, the WPAE also began inviting Southern donors to the table. However, the agenda was clearly that of the Northern donors. It included issues like ownership and accountability, transparency, predictability, and monitoring and evaluation. Thereby, the DAC sought to include the Southern development actors in its own existing framework.

These early efforts by the DAC to align Southern development actors to the norms of the DAC have been mirrored by other development actors. In 2008, for instance, the World Bank created the South-South Experience Exchange Facility, a special trust fund to finance knowledge sharing among development partners from the Global South as well as the Global North. The idea was to quickly disseminate experiences and results from SSC and thereby further cooperation (Abdenur and Da Fonseca 2013).

Likewise, a number of bi-and multilateral donors have pushed the TDC idea. In essence, TDC is a formalised North-South-South development relationship, which seeks to bring together the strengths of a Northern and a Southern donor to the benefit of a recipient country (see also Textbox 6.1). TDC is both a means to cooperate and to assimilate. In the words of Farias (2015: 2) TDC: '*can help "traditional" donors to strengthen their relationship with "emerging" donors, particularly those whose global influence transcends the realm of development assistance*'. Concurrently, it is a means to incrementally assimilate Southern development partners to the 'established' norms as the Northern partner often controls the resources, initiates the collaboration, and sets up the monitoring systems.

It is important to point out that the Global North's awareness of emerging changes in the international development cooperation arena not only gave rise to processes of cooperation and capturing vis-à-vis the development actors of the Global South, it also caused an internal power struggle among 'established' development actors. This for instance is seen in the efforts to create an alternative to the DAC, i.e. the United Nations Development Cooperation Forum (DCF), which was launched by the G77 as a '*more legitimate alternative to the DAC*' (Verschaeve and Orbie 2016: 572) and thus a way to counter what was perceived as a closed and exclusive forum. In short, the DCF was perceived as a more egalitarian development forum that could give voice to all development partners – from North and South alike, and private, public, and civil society organisation as well. The DCF was officially launched in July 2007 with a mandate to review trends in international development; promote coherence among development activities, and strengthen the work of the UN. Its role was perceived as being '*to provide a new window of opportunity for Southern countries to reflect their distinctive voices in development cooperation*' (Quadir 2013: 332), but due to organisational weaknesses and a focus on reviewing trends in international development rather than pushing for a more inclusive political voice, it has yet to provide SSC actors with a new platform to voice their views. Moreover, despite its indirect critique with the 'established' Northern dominated system, it has yet to outmanoeuvre the DAC.

Box 6.1 Trilateral development cooperation

TDC is often hyped as the new black in development cooperation. It is perceived as cheaper, more efficient, and more relevant than traditional (North-South) development cooperation. Therefore, it came as no surprise that it was '*listed as one of the eight "building blocks" of development at Busan* [4th High Level Forum on Aid Effectiveness]' (Masters 2014: 179), and by 2011, two-thirds of DAC donors had already engaged in TDC projects (Farias 2015).

TDC is not a novel idea. It can be traced back to the Buenos Aires Plan of Action (see chapter 2) that among other things gave priority to technical cooperation among developing countries, but lately it has been revived, not least in light of Millennium Development Goals (MDG) number 8, the global partnership for development. In essence, TDC is '*a development relationship in which a DAC donor and/or multilateral agency (e.g. Japan, Germany or the United Nations Development Programme) "partners" with a so-called "pivotal" country (e.g. Brazil, Thailand or South Africa) to work with a third "partner" (recipient) country (e.g. Ghana, Laos or Mozambique)*' (McEwan and Mawdsley 2012: 1186).

The idea is that the 'DAC donor' supplies finances and institutional capacity, while the 'pivotal' country brings to the table cheap goods and services, appropriate technology, and recent experiences of structural transformation. Thereby, TDC in theory brings together the best from North-South development cooperation with the best from SSC and the recipient country should ideally benefit from low costs, efficiency, appropriate technology, and cultural/geopolitical understandings.

Seen from the perspective of the 'DAC donor' TDC is useful as it potentially offers cheaper and more efficient development assistance; it is seen as a way to share responsibility for the challenges of poverty and inequality that the world is facing; and it provides the Northern donor with a platform to influence the norms of the 'pivotal' donor and thus blot out the existing differences. For the 'pivotal' country, TDC is perceived as a way to combine foreign policy, diplomacy, and development assistance while simultaneously easing the transformation from recipient to donor, i.e. bridging the role between developing and developed country (Masters 2014).

However, we know very little about whether this is indeed what happens. First, most of the information we have originates in donor organisations – not from independent analyses of the impacts of TDC. Second, the term TDC covers a multitude of activities involving not only state actors but also civil society organisations, philanthropists, and private sector actors. Moreover, it not only involves central state actors but also provincial governments and cities. Third, and related, 'DAC donors' perceive their role in TDC in different ways: whereas some 'dictate' the cooperation others put emphasis on mutual planning, implementation, and financing. Finally, TDC covers all sectors of the economy.

In theory, then, TDC is cheaper than North-South Cooperation due to lower salaries and cheaper technologies; it is more appropriate due to the use of intermediate technologies and policies; and it brings more funds to the development table. Simultaneously, it is also more complex, and legal procedures may be cumbersome; it is seldom based on equality; and it is unclear whether TDC is any better than North-South Cooperation in taking the recipients' needs into account.

Sources: (McEwan and Mawdsley 2012; Farias 2015; Masters 2014)

Discursive changes in the North: adopting the vocabulary of the South

The phase directly following the 'established' development partners' identification of a rejuvenation of SSC was characterised by a combination of courtesy and condemnation. Courtesy was unfolded via a mixture of formation of fora, roundtables, and conferences to further understanding and collaboration, whereas condemnation was disclosed in binary debates that opposed 'traditional' and 'emerging' development actors and by this means made the differences seem much bigger than the commonalities.

This phase was replaced after a few years by a phase characterised by an acceptance of the presence of the 'new' actors on the development scene; a replication of some of their modes of operation (focusing on infrastructure rather than social sectors and mixing modes of finance); and a reproduction of their discourse, i.e. how they portrayed their relationship with the rest of the Global South.

While this change obviously happened gradually, most commentators and researchers mention the fourth HLF in Busan, South Korea, in 2011, as the key tipping point in this development (Kim and Lee 2013; Mawdsley, Savage, and Kim 2014). No doubt, previous HLFs have also had a huge impact on development cooperation: the 2005 HLF in Paris, France, is a sponsor of the Paris Declaration focusing on ownership, alignment, harmonisation, mutual accountability, and results, and the 2008 HLF in Accra, Ghana, resulted in the signing of the Accra Agenda for Action, which on top of ownership and results, highlighted partnerships (with civil society organisations) and capacity development. While Southern development partners were signatories of the Paris Declaration (as recipients – not as donors), they were not really in focus in the Accra Agenda for Action, which paid particular attention to civil society organisations.

The Busan HLF was different. First, China and other Southern state actors actively participated in writing the documents. Second, the focus was on 'differential commitment', i.e. while Northern donors were obliged to follow the prescriptions of the Busan HLF, Southern donors could do it on a voluntary basis (Kim and Lee 2013). Hence, Western officials went some way to convincing Southern development partners to sign up to principles on aid effectiveness on condition that the special nature of SSC was recognised. The result was that the aid effectiveness debate was replaced by a broader development effectiveness paradigm that took the 'emerging' donors' development finance mechanisms into account (Kragelund 2015). There is no fixed definition of the development effectiveness paradigm but according to Mawdsley, Savage, and Kim (2014: 30), it:

> *includes a renewed focus on economic growth, enhancing industrial productivity and wealth creation (rather than poverty reduction per se); greater integration between foreign aid and other policy areas, such as trade, investment and migration; and a growing and more visible role for the private sector.*

Meanwhile, the MDGs, adopted in late 2000 by close to 200 countries, were coming to an end. In short, the MDGs were eight measurable goals, including halving poverty, increasing primary education, promoting gender equality, reducing child mortality, ensuring environmental sustainability, and developing a global partnership for development, that were to be met by 2015. More importantly, however, the period from 2000 to 2015 *'represented an unprecedented period of international agreement about what 'development' consists of, and clear targets as to what was to be achieved'* (Willis 2016: 1).

By the end of this period, this agreement came to an end. Not only were the MDGs only partially met, the development community also began realising that development was more than poverty alleviation. The formulation of a new set of goals, the Sustainable Development Goals (SDGs), to succeed the MDGs were set in motion in 2012 and approved by the General Assembly of the UN in September 2015. Essentially, the spatial coverage, ambition, and complexity of the SDGs were much higher than the MDGs. Not only were the number of goals increased from eight to 17, the SDGs also included no less than 169 targets. Moreover, the SDGs were no longer limited to developing countries, but acknowledged that challenges relating to housing, gender, and the environment are as widespread in the Global North as they are in the Global South, i.e. the goals were for all – not only the poor and destitute in the Global South. The spatial coverage, ambition, and complexity of the SDGs imply that the financial requirements are even higher than the MDGs. Even though 'traditional' development partners and ODA will contribute to the SDGs, the financial framework is much broader than that. It includes Southern development partners, trade, investment, and debt forgiveness/rescheduling (Willis 2016).

What we see therefore is not only a rhetorical break with development being equivalent to poverty alleviation in the Global South; the SDGs also point towards '*a shared mission as well as a mutual interest*' (Keijzer and Lundsgaarde 2017: 4), which legitimises national development partners to consider a shift towards mutual benefit as well.

Reactions from individual DAC members

The changes that we have witnessed in the DAC towards a 'Southernisation' of development finance have been mirrored at the individual country level. Due to a combination of economic downturn after the global financial crisis, the election of populist right-wing governments across the Global North, growing aid fatigue, and a realisation that 'emerging' economies indeed compete with the 'traditional' partners for both business opportunities and 'hearts and minds' in the Global South, many DAC donors have recently changed their discourse on aid. Whereas DAC donors in the 1990s and the first decade of this millennium officially cast their aid programmes in terms of 'doing good', i.e. poverty alleviation, improved primary health, primary education, (gender) equality etc., the discourse today is centred around national self-interest, value-for-money, and win-win (Mawdsley 2017b, 2015), see Table 6.1.

This is particularly pronounced in the case of the United Kingdom where the Conservative party in 2009 adopted a new 'Green Paper' on international development that pointed towards 'value-for-money' and mutual benefit. Soon thereafter British aid became '*subject to rigorous tests of value-for-money and effectiveness*' to make sure, among other things, that it also would '*bring benefits to the UK*' (Mawdsley 2011). Since then, British aid has adopted a discourse that is even

Table 6.1 *Development policy statements of select DAC members*

United Kingdom (2015)	Netherlands (2013)	Denmark (2016)
'The UK's development spending will meet our moral obligation to the world's poorest and also support our national interest'	'Our mission is to combine aid and trade activities to our mutual benefit'	'We will be driven by the wish to promote Danish foreign and domestic policy interests at one and the same time'

Source: Adapted from (Keijzer and Lundsgaarde 2017: Box 1)

more explicit with regard to national interest. This national interest, however, is not only economic, i.e. national firms benefitting from the development finance directly or indirectly, but also relates to soft power (the British aid agency, DfID, receives international acclaim for its work and often takes the lead in international negotiations); global public goods, e.g. environmental sustainability, climate change mitigation, and disease control; and security (Mawdsley 2017b).

Dutch aid is also moving from focusing mainly on poverty alleviation to directing aid towards economic growth (Breman 2011), and so is the representation of Danish aid. Until recently, it was portrayed as being driven primarily by moral concerns (Lancaster 2008), but this is changing now. Now, the derived benefits for the Danish private sector are the key argument not to cut spending on ODA development aid (Kragelund 2015).

What we see therefore is that bilateral DAC donors now present their development aid in terms of national self-interest, win-win, and mutual benefit – exactly like Southern development partners. DAC donors are thereby repositioning development in the context of a perceived need to justify aid vis-à-vis other priorities.

Reactions from the IFIs: conditionality, debt sustainability, and voting power reform

One of the biggest controversies in the SSC debate revolves around the effects on political and economic developments in the rest of the Global South; that is, to what extent the rejuvenation of Southern development partners – and in particular China – would lead to (liberal) political and economic retrogression due to lack of conditionalities attached to development finance from the Southern development partners (Woods 2008), see also chapter 7. In order to further our understanding on this issue, we have to step back a bit, namely to how the IFIs have reacted to the growing presence of these 'emerging' donors who provide development finance with 'no strings attached'. In principle, the IFIs have two options: either they impose more/stricter conditionalities to their loans to counteract the lack of requirements from actors like China and Saudi Arabia and thereby ensure that the recipient countries change their policies in the direction the IFIs want, or they impose fewer conditions to make their loans attractive to recipient countries.

History has taught us that the latter is most likely the case. Until the debt crisis of the 1980s (see chapter 2) and the subsequent structural adjustment and stabilisation programmes imposed by the World Bank and the IMF, respectively, these IFIs *'lacked the leverage to negotiate agreements with recipient country governments'* (Hernandez 2017: 530). The demand for their loans was simply too small to give them any clout. The availability of development finance from a plethora of actors – including the Southern development partners – may similarly lower the IFIs' bargaining power as alternative resources are available, cf. (Whitfield 2008).

According to Hernandez (2017), this is exactly what has happened for the World Bank. He has analysed the number of conditions attached to loans to Africa from 1980–2013 and finds that on average the World Bank provides 15 per cent fewer conditions to loans to African governments if they are also assisted by China. This figure is even higher for the least developed African countries. Similarly, the World Bank attaches fewer conditions to middle-income countries in Africa also assisted by Kuwait and UAE. The World Bank is hence changing its lending policies as a result of SSC.

Another way to approach the controversy is to examine how or the extent to which IFIs have altered fundamental norms to accommodate the growing competition from Southern development finance providers. One such norm is debt sustainability. The IFIs perceive it as prudent debt management; that is, the country's repayment ability via a revenue model. In contrast, China perceives it as development sustainability; that is, the ability to repay the loans in the long term via economic development.

Data on norm changes is still scarce but a study of the controversies over the conditions of a large-scale resources-for-infrastructure loan by China in the Democratic Republic of the Congo (DRC) may shed some light on how IFI norms are affected (Malm 2016). In short, President Kabila needed large-scale infrastructure investments in order to retain popularity among voters and to increase his chances of re-election in the 2006 Presidential elections. An IMF loan programme had recently been terminated and, due to the IMF's position as gatekeeper, the DRC was not in a position to access finance from its 'traditional' donors. Kabila, therefore, turned to China, which was interested in the resource-rich Katanga Province. The result was the

Sicomines agreement, a joint venture of two Chinese companies and the Congolese state, funded by non-concessional loans from the ExIm Bank of China and repaid via profits from the mine. The agreement and credit lines worth US$6 billion would give the joint venture access to copper and cobalt in return for finance for infrastructure.

A loan this size would increase the debt burden dramatically for the DRC and hence, DRC's 'traditional' donors united behind an agenda to reduce the size of the loans. Similarly, they were displeased with the Congolese state's sovereign guarantee for the entire loan, i.e. a pledge of total reimbursement if the profits from the mining operations were insufficient to cover the loans. In the words of Jansson (2013: 155): '*The IMF argued that this guarantee was problematic from a debt sustainability perspective and was unreasonable, since no other investor in the DRC's mining sector has a government guarantee on the return on its investment*'.

The size and conditions of the credit lines became the focal point of an intense arm wrestling between China, the IMF, and the Congolese state. In the end, however, the size of the loans was reduced and conditions resembling IFI conditions were applied. In other words, the IMF did not change its norms to become competitive in the DRC (Kragelund 2015; Jansson 2013).

Finally, one could analyse the IFIs' reaction through the lens of IFI governance structures. The tectonic shifts in global economic power over the last couple of decades have caused a heated debate over the governance structure in the IFIs. In short, emerging economies as well as developing countries reasoned that the existing voting system did not reflect the current distribution of economic power in the world. The 2007/8 global financial crisis only underlined the importance of speedy reforms in the IFIs.

After a long run-up the IMF and the World Bank Group announced the so-called 'voice reforms' in 2010. In essence, these reforms should give emerging and developing countries more voice in the IFIs – and consequently reduce the voting power of the developed country members. The stated aim was to make the current size of the economy determine voting power more directly than the pre-reform period where voting power was largely determined by the size of the economies at the time of the creation of the Bretton Woods institutions.

Although the changes do not in the slightest reflect the current distribution of economic power in the world, realignment of voting power did take place. Both low-income and middle-income countries managed to increase their share of the total votes at the expense of the high-income countries (see Table 6.2).

However, these relative changes conceal both the minuscule absolute changes that took place in the World Bank (in percentage points voting rights of low-income countries only increased by 0.39 while middle-income countries' voting rights increased by 3.32 percentage points) and the inequity of the voting system. As stated above, a major aim of the voice reforms was to align voting rights more to the current size of the member states' economies, but even if China has experienced an increase of 1.64 percentage points and Japan's voting power has decreased by 1.01 percentage points (see Table 6.3), the share of voting power to GDP is by no means even.

Table 6.2 *Voting power realignment in the World Bank (shareholding %)*

	BEFORE	*AFTER*	*CHANGE (%)*
LOW-INCOME COUNTRIES	3.45	3.84	11.30%
MIDDLE-INCOME COUNTRIES	31.22	34.54	10.63%
HIGH-INCOME COUNTRIES	65.33	61.62	−5.68%

Source: adapted from (Vestergaard and Wade 2013: Table 2)

Table 6.3 *Results of voting power reforms in the World Bank*

Increase (percentage points)		*Decrease (percentage points)*	
China	1.64	**Japan**	−1.01
South Korea	0.58	**France**	−0.55
Turkey	0.55	**United Kingdom**	−0.55
Mexico	0.50	**United States**	−0.51
Singapore	0.24	**Germany**	−0.48
Greece	0.21	**Canada**	−0.35
Brazil	0.17	**Netherlands**	−0.29
India	0.13	**Belgium**	−0.23

Source: adapted from (Vestergaard and Wade 2013: Table 1)

If voting power fully reflected the size of the economy, the voting power to GDP ratio (VP/GDP) would be 1. However, as is apparent from Table 6.4, this is not the case. While Saudi Arabia's VP/GDP in the World Bank is close to 4, China's VP/GDP is only 0.43, i.e. Saudi Arabia's VP/GDP is nine times higher than China's VP/GDP. Similarly, Belgium's VP/GDP is three times higher than India's VP/GDP.

The discrepancies in VP/GDP are not as severe in the IMF compared to the World Bank. The voice reform in the IMF is still to be accepted by more than 85 per cent of the votes, but if the amendments are agreed upon, China will still only end up with one-fifth of the VP/GDP compared to, for example, Belgium.

As described above, the IFI's reaction to the rejuvenation of SSC is ambiguous. On the one hand, there is no doubt that the World Bank imposes fewer conditions to countries also courted by China and the Arab donors. On the other hand, there are indications that the IFI are not (yet) willing to compromise on their norms and values. The result

Table 6.4 *Post 'voice reform' voting power (VP) to GDP rations in the World Bank and IMF, respectively*

World Bank				IMF[a]			
Member	*Share of GDP (%)[b]*	*Share of VP (%)*	*VP to GDP ratio*	*Member*	*Share of GDP (%)[c]*	*Share of VP (%)*	*VP to GDP ratio*
United States	22.29	15.85	0.71	**United States**	20.53	16.47	0.80
China	10.37	4.42	0.43	**China**	13.63	6.07	0.45
Japan	7.34	6.84	0.93	**Japan**	6.84	6.14	0.90
India	3.60	2.91	0.81	**India**	4.34	2.63	0.60
Brazil	2.74	2.24	0.82	**United Kingdom**	3.12	4.02	1.29
Netherlands	1.16	1.92	1.65	**Brazil**	3.02	2.22	0.73
Saudi Arabia	0.72	2.77	3.86	**Netherlands**	0.98	1.76	1.80
Belgium	0.68	1.57	2.30	**Belgium**	0.61	1.30	2.14

Notes: [a]) Two years after the decision to put the IMF voice reform into action, less than 50 per cent of the votes had accepted the amendments. It takes 85 per cent of the votes to amend the voting power in the IMF;
[b]) 2009 GDP data based on a (60/40) blend of current GDP (60 per cent) and GDP purchasing power parity (40 per cent);
[c]) 2012 GDP data based on a (50/50) blend.

Source: adapted from (Vestergaard and Wade 2015: Table 1 and 2; 2013: Table 3).

of the governance reform of the IFIs is also vague: no doubt, power has been transferred from high-income countries to middle-and low-income countries, but they appear much more substantial than they really are (Vestergaard and Wade 2013) and as this section has shown the IFIs are still a far cry from letting voting power reflect economic muscles.

Dynamics of the relationship: homogenising and differentiating processes

The preceding sections have documented how discourses, norms, and practices have changed during the past decade and now increasingly mirror the discourses, norms, and practices of the Global South – what Emma Mawdsley recently has referred to as the 'Southernisation' of development cooperation. This, however, is not the only change that has taken place. As I have argued elsewhere, we currently witness a process convergence between the Global North and the Global South (Kragelund 2015).

Chinese development cooperation is an illustrative case in terms of understanding the processes that are currently taking place in the Global South. Chinese aid has often been portrayed as *the* rogue donor (Naím 2007) and often the focus is on stable principles and policies (Li 2007; Zhou 2012) rather than its sophisticated nature, its complexities, and the changes taking place.

Table 6.5 presents a stylised picture of key characteristics of China's development finance to Africa from its onset in 1956 to the most recent FOCAC held in Johannesburg in 2015. It demonstrates that while the overall aim of this relationship has been relatively stable, both emphasis and focus have changed incrementally and followed, with a time lag, the trends of the Global North. For example, global warming has gradually come to play a more and more important role in Sino-African relations, first under the title of 'environmental protection' during FOCAC III, then named 'climate change' in FOCAC IV but only receiving little attention, and finally moving to a prominent position during FOCAC V. Likewise, the focus on infrastructure has changed from a strictly local focus to a more regional and continental focus linking different parts of Africa together, and ending in FOCAC VI with a focus on connecting Asia with Africa, Oceania, and Europe via the 'Maritime silk road', a section of Chinese development strategy

Table 6.5 Selected characteristics of Chinese development finance to Africa, 1956–2015

	1956–1978	1979–1989	1990–1999	FOCAC I 2000	FOCAC II 2003	FOCAC III 2006	FOCAC IV 2009	FOCAC V 2012	FOCAC VI 2015
AIM	Diplomacy	Diplomacy & eco. motives	Int. politics & eco. motives	Diplomacy and economic cooperation					
DRIVING FACTORS	Ideology and geopolitics	Eco. restructuring in China	Geopolitics and promotion of Chinese exports	New world order	Safeguarding common interests	Coordination of positions in relation to global governance			
MODALITY	Economic aid	Grants and loans (towards more concessional loans – especially after the 1995 reform of aid)		Grants, concessional loans, and debt cancellation	Grants and concessional loans	Specific pledges in relation to aid, loans, and debt cancellation	Grants, concessional loans plus debt cancellation	Grants and concessional loans	Grants, concessional loans, trade, and investment
EMPHASIS	Solidarity, mutual benefit, and equality	Efficiency and mutual benefit	Link aid, trade, and investment	Common prosperity	Win-win cooperation		Global recession	Climate change	Win-win cooperation

(Continued)

Table 6.5 *(Continued)*

	1956–1978	1979–1989	1990–1999	*FOCAC I* 2000	*FOCAC II* 2003	*FOCAC III* 2006	*FOCAC IV* 2009	*FOCAC V* 2012	*FOCAC VI* 2015
MAIN FOCUS			Capacity building, technical training	Trade and commerce, local industries, local materials, and job creation	Economic development, especially agriculture & local infrastructure	Agriculture and infrastructure	Business, agriculture and food security, infrastructure, and climate change	Climate change, peace and security, agriculture, and infrastructure	Infrastructure, agriculture, industrialisation, natural resources, tourism, and Maritime Silk Road
SECONDARY FOCUS				Health, edu., and human resources	HRD, health, and cultural exchanges	Counter terrorism, human resources, health, and environmental protection	Health, edu., science and tech.	Health, edu., and academic exchanges	Humanitarian assistance, health, edu., poverty eradication, & climate change, anti-terrorism

Source: Adapted from (Kragelund 2015: Table 1)

to connect China to Eurasia, South East Asia, and South Asia, the so-called 'One Belt One Road' initiative.

In the same vein, softer issues such as health, education, and personal exchange are receiving ever more attention. New on the agenda, too, are security concerns, which are linked to the rapidly growing Chinese investments all over Africa, and poverty eradication.

Overall, it can be seen that the main aim is no longer diplomacy and international politics but rather economic development. Also, the main drivers have changed from internal restructuring in China to issues related to global governance. Similarly, modalities, emphasis, and sectoral focus have changed incrementally. Now, Chinese development finance pays more attention to topical issues and gives more priority to social sectors. This resembles the activities of the Global North over the past couple of decades (Kragelund 2015).

We can also detect a move towards more transparency in the development cooperation activities of China, including: miscellaneous news on projects from the Ministry of Commerce, a publication in 2011 of China's foreign aid, the so-called 'white paper' on China's aid, and a Foreign Aid Communication bulletin. The move towards more transparency has been mirrored in the semantic debate. Whereas China of late has deliberately used the term 'external assistance' in policy papers in order to distance itself from the Global North, it now uses the term 'aid' like the DAC donors.

Conclusion

This chapter has revealed a number of changes to development 'as it was'. It has shown how the 'established' donor community first reacted with fear and then tried to simultaneously learn from the 'new' development actors and assimilate them into the DAC line of thinking. The changes that took place to 'development as it was' in this early phase were not confined to the encounter between Northern and Southern development providers, it also included a clash between the DAC and the UN in terms of who were to define future development cooperation.

The first phase only lasted five to six years whereupon it was replaced by a phase characterised by discursive changes in the North, i.e. what Emma Mawdsley has called a partial 'Southernisation' of development cooperation. The Busan HLF in 2011 marks the tipping point in this phase. Here, the Global North officially adopted the vocabulary of the Global South and bi-and multilateral donors alike openly acknowledged that the purpose of aid no longer solely is social and economic development in the Global South, but also has to benefit the domestic private sector in the Global North.

In parallel with these changes the IFIs have met the increasing competition from providers of development finance from the Global South by demanding fewer conditions and by incrementally changing their governance structures.

These changes are not only caused by the rejuvenation of SCC. SSC is only one – but an important – part of a move towards more complexity, more actors, and less predictability in international development cooperation. No doubt, many of the reactions we have witnessed in the Global North over the past decade are attributable to long-term changes in global economic power, the global financial crisis, and a populist turn in domestic politics in the Global North. Equally important is a growing aid fatigue; a proliferation of development actors beyond Southern state actors, such as businesses, super rich individuals/foundations, new EU actors; innovative sources of finance; an increasing focus on global public goods; and the diminishing importance of ODA vis-à-vis other flows of development finance.

Finally, this chapter has argued that the changes that we are currently witnessing are not only the result of changes in the Global North, but also changes in the Global South. It has described how Chinese development cooperation is in no way static but rather converges towards development cooperation of the Global North which is itself also changing.

Discussion questions

- Describe how the 'established' donors reacted to the rejuvenation of SSC.
- Explain in particular how and why their response to SSC changed over time.
- Discuss the extent to which changes in the IFIs as a result of the revitalisation of SCC change the power relations between the Global North and the Global South.

- Discuss how the definition of debt sustainability matters for the recipient countries.

Web pages of interest

- The China-DAC study group was formed in 2009 with the aim of sharing knowledge and exchanging experiences between China and the DAC on promoting growth and reducing poverty in the Global South. The web page contains information on the group and its activities in the past: www.oecd.org/dac/dac-global-relations/china-dac-study-group.htm
- The official UN web page for the DCF contains information on its rationale, what is has done and what it plans to do in the future: www.un.org/ecosoc/en/development-cooperation-forum
- OECD's web page on its past four HLFs on Aid Effectiveness includes information of the outcomes of each forum: www.oecd.org/dac/effectiveness/thehighlevelforaonaideffectivenessahistory.htm
- The official web page for the Millennium Development Goals that influenced the Global North's development agenda in the first one and half decades of the new millennium contains valuable information of what has been achieved by whom and for whom: www.un.org/millenniumgoals
- The SDGs succeeded the Millennium Development Goals in 2016. The official web page for the SDGs contains descriptions of all the goals: sustainabledevelopment.un.org/?menu=1300

Note

1 A number of different terms are used to describe the same phenomenon. They include 'triangular cooperation', 'trilateral assistance' and 'tripartite agreement' (Masters 2014).

Further reading

Manning, R. 2006. Will 'Emerging Donors' Change the Face of International Co-operation? *Development Policy Review* 24 (4):371–385. The first official acknowledgement that two decades of absolute hegemony in the development area were coming to an end.

McEwan, C., and E. Mawdsley. 2012. Trilateral Development Cooperation: Power and Politics in Emerging Aid Relationships. *Development and Change* 43 (6):1185–1209. Offers a critical analysis of what is often hyped as the new solution to aid inefficiency in development cooperation. The article examines trilateral development cooperation and concludes that while it may lead to more effective aid, it may also further depoliticise it and thereby questions whether this is the right route for SSC.

Vestergaard, J., and R. H. Wade. 2013. Protecting Power: How Western States Retain The Dominant Voice in The World Bank's Governance. *World Development* 46:153–164. This article offers a systematic analysis of the changes in the global governance structure that took place in the World Bank after the global financial crisis. It scrutinises the mismatch between rhetoric and realities.

Woods, N. 2008. Whose aid? Whose influence? China, emerging donors and the silent revolution in development assistance. *International Affairs* 84 (6):1205–1221. Influential response, to among other things, Richard Manning's fear that the rejuvenation of SSC would lead to increasing debt burdens, postponement of much needed reforms, and scarce use of productive resources. Woods shows how development aid has been quite unsuccessful in driving changes in the Global South and that the 'emerging' donors are not as bad as was often portrayed in the early years of SSC research.

7 Effects of South-South Cooperation on the rest of the Global South

Studying local effects of SSC: an analytical framework

Ever since the interest in SSC rejuvenated a bit more than a decade ago academics, politicians, and journalists have been interested in understanding how and to what extent the growth of this phenomenon affected the rest of the Global South. Regrettably, some conclusions are rather normative and emotive and have been drawn on a combination of poor data (see also chapter 1) and/or superficial reading of available information. Other studies seem to group in two ends of a continuum, either praising SSC (or certain vectors of it) (Sun 2017; King 2013) or vilifying it (Tull 2006; Eisenman and Kurlantzick 2006). Only recently, more nuanced studies of opposing effects of SSC for host economies have come to the fore (Carmody 2013; Mawdsley 2012b; Chaturvedi, Fues, and Sidiropoulos 2012; Large and Patey 2011).

A group of scholars centred around the Institute of Development Studies, Sussex, were among the first to propose an analytical framework to understand how 'Asian drivers' affected the developing world (Kaplinsky and Messner 2008). In short, they distinguished between whether the impacts of for instance China's engagement in a particular African country would be complementary to the economic activities already taking place in that country – whether driven by actors of that country or third country actors – and thus form a 'win-win' situation, or they would be competitive and thus form a 'win-lose' situation. Moreover, they distinguished between direct and indirect impacts, i.e. between impacts that follow directly from for instance India's export of labour-intensive capital equipment to Kenya's food processing sector allowing Kenyan entrepreneurs to produce more cheaply and effectively (complementary direct impacts) and impacts that result more indirectly from this relationship, for instance China's exports of motorbikes to Ghana that replaces Nigeria's export of motorbikes to Ghana (competitive indirect effects). Finally, they argued that this engagement could simultaneously result in different sorts of complementary/competitive and direct/indirect effects and that it is dynamic.

Figure 7.1 presents this framework with regard to trade. It shows how SSC trade can lead to both win-lose and win-win outcomes and how this can happen directly as well as indirectly. Farooki and Kaplinsky (2012) provide examples of how China's trade can lead to these effects. In the lower left quadrant of the figure are the direct complementary impacts, which may derive from exports of for instance low-price Chinese consumer products that increase welfare for less affluent consumers in the Global South. The direct complementary trade impacts also include increased imports of primary commodities from for instance resource-rich African

Figure 7.1 *Economic impacts of South-South trade on the rest of the Global South*

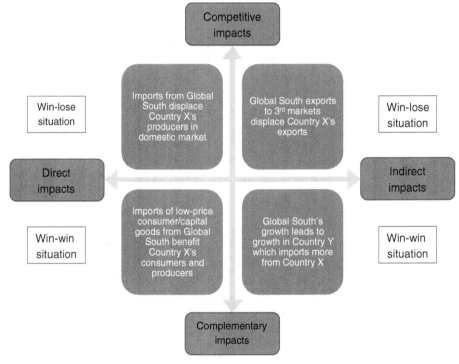

Source: Adapted from Farooki and Kaplinsky (2012: Figure 2.4)

economies to China. All SSC trade, however, is not beneficial to the rest of the Global South. The upper left quadrant of the figure points to the competitive direct impacts of increased SSC trade. The Ethiopian shoe and leather industry a decade ago is a classic case of how Chinese imports (at first) outcompeted local producers (see chapter 5), but these impacts are not confined to Chinese exports. In Zambia, South African products fill up the lion's share of the shelves in all major supermarkets in Zambia (Kragelund and Carmody 2016). The upper right quadrant points towards the indirect competitive impacts of South-South trade. The textile and clothing sector is a classic example of these type of impacts: the end of the Multi-Fibre Agreement in 2004 meant that Chinese producers of textiles and clothes could freely export their products. As they were significantly more efficient than their African counterparts, the end of the Multi-Fibre Agreement meant that for instance Kenyan producers were unable to compete internationally (McCormick, Kamau, and Ligulu 2006), see also chapter 3. Finally, the lower right quadrant highlights the indirect complementary effects of South-South trade. Two examples come to the fore here: first, China's growing demand for primary commodities leads to higher prices even though China is not buying commodities from a particular country. Second, the demand for commodities has driven economic growth in a number of resource-rich economies in the Global South that has led to demand for goods and services in third countries.

This analytical framework is applicable to all the vectors of engagement. Table 7.1 lists examples of direct complementary and competitive effects of SSC in the vectors described in chapter 4. Development finance, for instance, may add much needed finance to both public and private actors in the Global South. Likewise, it targets sectors not targeted for much of the past two decades by the Global North, and development finance from the Global South uses other modalities than development finance from the Global North. In contrast, subsidised development finance may displace local financial intermediaries and thus retard the development of sustainable local financial institutions.

Likewise, South-South investments may have conflicting direct impacts on the rest of the Global South. FDI by definition is a source of finance, but the sheer amount of FDI does not inform us about

Table 7.1 *Examples of direct competitive and complementary impacts of SSC*

Vectors	Impacts	Nature of link
Development finance	Complementary	Additional ODA-and OOF-like flows to state and private actors
	Competitive	Low-cost development finance displaces local financial intermediaries
Trade	Complementary	Imports of cheap consumer goods and capital equipment, and exports of commodities
	Competitive	Imports of consumer goods displace local production
Investments	Complementary	Inflows of FDI
	Competitive	Competition with domestic firms
Migration	Complementary	Migrants from Global South intermediate complementary trade with home economies
	Competitive	Migrants from Global South displace local entrepreneurs
Education	Complementary	Additional possibilities for education and further training
	Competitive	Subsidised educational programmes may displace local programmes
Global governance	Complementary	Support for a development agenda in international fora
	Competitive	Parts of Global South side with Global North in international fora

Source: Adapted from Kaplinsky and Messner (2008: Figure 6)

the developmental effects of the investments. What matters is what technologies are used (and to what extent technology transfer results in capability building); how many people are employed (and under what conditions); whether the foreign firm sources supplies and services locally; whether they serve an export or a domestic market; to what extent profits are repatriated; and whether (and how much) tax is paid in the host economy. Moreover, it is important (initially) to distinguish between so-called green field investments, i.e. investments leading to a new venture in the host economy, and brown field investments, i.e. purchase of an existing company. While the latter by definition replaces an existing company (sometimes locally owned), the former by definition adds new economic activities in the country.

South-South migration may also lead to both positive and negative outcomes for the host country. On the one hand, migrants bring along new skills and new products that benefit the host population. On the other hand, migrant entrepreneurs may outcompete local entrepreneurs either forcing them out of work or reducing profit margins. South-South education programmes may also result in contradictory outcomes: while they may lead to more people being offered free or highly subsidised education or further training these programmes seldom follow standards set locally and they may outmatch local programmes that have to work on a for-profit basis.

Finally, the direct effects of South-South Collaboration in global governance are also ambiguous. As depicted in chapter 5, collaborations like IBSA pushed a development agenda in international fora like the WTO but this vector is also the prime example of the heterogeneity of the Global South: often some countries of the Global South align with the Global North while the majority of countries stick together.

The analytical framework (Figure 7.1) thus provides us with an idea of where to look for direct and indirect effects of SSC and Table 7.1 presents us with examples of the former in the Global South for the six vectors of engagement presented in chapter 5. However, both the framework and the examples are less good at informing us how the Global South is affected differently by SSC and how and to what extent time matters for the effects that we see. The following three sections will seek to uncover this complexity by examining in more detail the economic, political, and social effects, respectively, of SSC.

Economic effects

Even if a decade of research into SSC has demonstrated that the relations between the economies of the Global South are highly complex; that the Global South is incredibly heterogeneous; and that all the vectors of engagement may lead to both win-win and win-lose situations, a substantial part of the contributions to SSC are surprisingly one-sided in their conclusions.

A recent study based on visits to approximately 50 Chinese-owned factories in sub-Saharan Africa presents the following conclusion in the opening pages: '*Chinese factories in Africa: This is the future that will create broad-based prosperity for Africans and usher in the next phase of global growth for a large swath of the Chinese economy,*' and continues, '*Factories are the bridge that connects China, the current Factory of the World, to Africa, the next Factory of the World*' (Sun 2017: 6). Presented this way, SSC, and in particular 'China in Africa', thus automatically leads to a win-win situation for the 'emerging South' as well as for the 'Rest of the South'. Admittedly, Sun (2017) points to some of the environmental and social downsides of Chinese investment in Africa, but argues that these negative aspects are 'somehow necessary' as they lead to better regulation and that historically regulation has always come after industrialisation – not before.

This argument largely follows former chief economist of the World Bank Justin Yifu Lin's optimistic predictions of how economic growth in China, in particular, may lead to structural transformation in African economies (Chandra, Lin, and Wang 2013; Lin 2012). The argument is that as labour reserves are absorbed in China, the country will move from labour-intensive towards more capital-intensive production. In the process, labour-intensive manufacturing jobs will move towards other world regions with lower wages – including Africa.

Lin (2012) builds his predictions on the so-called 'flying geese' concept originally developed in the 1930s by Akamatsu Kaname. The flying geese metaphor refers to the inverse V-shape formed by migrating geese and was originally used to explain the spatial pattern of East Asian and South East Asian growth. The original model depicts the sequencing of import, production, and export and describes how 'poor' countries first import consumer goods, then begin production of these at the same time as they initiate the import of capital equipment, and finally begin exporting consumer goods, which stimulates the production of capital equipment. Each type of good undergoes a process from import, via production, to exports, and the real value of each of these processes in the 'poor country' first increases then decreases – forming an inverse V (Kojima 2000).

Lin (2012) and Sun (2017) apply the model somewhat differently. They use it to examine the relocation of old industries in the Global South to countries with lower production costs. Lin (2012: 405), for instance, argues that due to the sheer size of the Chinese economy,

'*China will not be a goose in the traditional leader-follower pattern of industrialisation for a few lower income countries but a dragon*'. In a similar vein, Chandra, Lin, and Wang (2013: 77) state that: '*It seems reasonable to suggest that the leading dragon phenomenon alone can create sufficient labour-intensive manufacturing jobs for developing sub-Saharan African countries to bring them to par with most industrial countries*'.

Empirical studies are yet to support this one-sided representation of the economic effects of SSC. In contrast, a recent study of Indian, Chinese, and South African investments in the mining and tourism sectors in Zambia concluded that Lin's:

> *analogy of China being a dragon that will structurally transform African economies is most likely wrong – at least in the Zambian case where . . . these investments to a large extent follow a similar path to investments from the Global North, i.e. with limited linkages and spill overs to local companies.*
>
> (Kragelund and Carmody 2016: 234)

A major reason for this discrepancy between the potential effects and what we hitherto have been able to see on the ground in Africa may relate to the composition of the economies of Africa. According to Taylor (2014: 128) half of the African countries derive more than 80 per cent of their export income from commodities and '*the upsurge of interest in Africa by BRICS and other emerging economies has coincided with – and possibly exacerbated – the continent's increased dependency on primary products, particularly mineral products*'. The result is that '*since the upsurge of interest in Africa by the BRICS . . . there have been very few signs of social transformation in Africa and in fact there have been signs of* deindustrialisation' (Taylor 2014: 131, emphasis in original).

The future, however, may change this picture incrementally for some African economies. Rising costs of manufacturing production in China, due to rising real wages, stiffer regulations, increasing coordination, and transport costs; reorientation of local production to cater for rising local demand in China and the rest of Asia; technological upgrading across manufacturing firms in Asia; and strategic political decisions to 'off-shore' Chinese production to Africa may according to Newman et al. (2016) pave the way for

Plate 7.1 *China's role in African industrialisation*

Source: (China's Industrial Colonialism in Africa © 2014 Zapiro. Originally published in The Times. Re-Published with permission – For more Zapiro cartoons visit www.zapiro.com)

industrialisation in Africa based on so-called 'task-based' trade, i.e. linking up to lead firms in global value chains (see chapter 4).

Despite the rising costs, stiffer regulation, and a political will to outsource production, this change is probably not just around the corner. The lion's share of Chinese jewellery, for example, that enters the African market is produced in the provinces of Guangdong and Zhejiang, China, by small-scale entrepreneurs with limited technological capabilities and limited access to capital. The Guangdong provincial government is in line with the central Chinese decision to offshore this kind of production to low-cost countries and keep high value-added production of fashion jewellery within its boundaries. Notwithstanding this verdict, unlicensed workshops producing low-grade jewellery for the markets in Guangzhou, the centre of China-Africa trade (see chapter 5), dominate production (Haugen 2018).

Chinese traders used to sell it in the markets of African cities, but recently local regulations have shifted this trade into local hands. In Ghana, for instance, there is a ban on foreigners bringing in more than one suitcase weighing up to 32 kilogrammes while Ghanaians are allowed to check in an additional 12 suitcases (for a fee). Likewise, Chinese entrepreneurs used to use Ghanaians as fronts in their jewellery shops in cities like Accra and Tema. This has also been stopped by local authorities. Essentially, this means that local trade of cheap jewellery is now in local hands. However, this does not entail that profit margins (anywhere in the chain) are big. As depicted in Figure 7.2, the price for a pair of standard cheap plastic earrings produced in Guangdong increased 32 per cent when sold by the Chinese supplier in 'Chocolate city'. The Ghanaian trader going to China to buy the earrings adds another 43 per cent to the price before the Ghanaian retailers selling them individually on the streets of Accra and Tema add another 70 per cent to the price.

Even where entry barriers are low both technology-and capital-wise and profit margins are low all along the chain, like in the cheap jewellery value chain from China to West Africa, there is no evidence yet that manufacturing production is relocating from the coastal areas of China to Africa thereby driving a process of structural transformation in Africa. Rather, the jewellery value chain provides a *'means for people to make a living in volatile times and in the absence of other opportunities. Informal commercial importers have effectively*

Figure 7.2   The Chinese plastic earrings to the Ghanaian market value chain

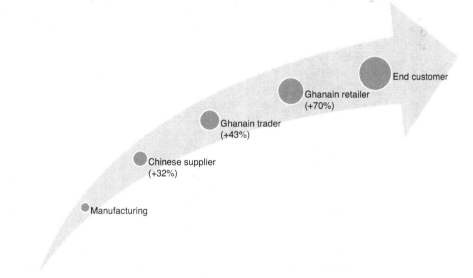

Source: adapted from Haugen (2018: Figure 1)

addressed supply shortages and provided people with affordable goods' (Haugen 2018: 322).

Aggregate data on change in trade and production structures may inform our general understanding of the economic effects of SCC. Similarly, analyses of push and pull factors for Chinese enterprises setting up factories in African economies may improve our understanding of which direction development will move in the years to come, but neither informs us who benefits and who loses out from the new trade and investment relations between countries in the Global South. Textbox 7.1 provides a snapshot of how SSC may lead to changing local power figurations. Here, Chinese-produced wax prints lead to cheaper prices and smaller units – a democratisation of the market – but also the demise of a powerful group of traders that benefitted from their close networks to the European producers.

Box 7.1 Chinese textiles and the diminishing power of West African 'Nana-Benzes'

Throughout West Africa 'local' wax print fabrics make up a large share of the local textile markets. The fabrics are used for ceremonies as well as for casual dress and they come in a variety of qualities and designs. In reality, the 'local' wax print has not been local for a long time. Rather, European companies monopolised the design and production of what is now known as local fabrics more than 150 years ago. This, however, does not entail that the West African textiles did not generate economic activities locally: European companies, like the Dutch-owned Vlisco, established production facilities in major cities of West Africa, and large-scale trading companies distributed the textiles to local agents who brought the wax print to the local markets.

For the past 70–80 years, these agents have played an important role in shaping the local market by creating oligopolistic market structures; by acting as trend spotters; by changing designs regularly; and by branding the 'European' wax print as high-rank. This position soon resulted in high profits that allowed these women agents to buy Mercedes Benzes, which no one else could afford. Hence, they were soon nicknamed 'Nana-Benzes'; Nana in West Africa is used both as the title of a monarch as well as a way to denote social eminence and age; and Benz is the shorthand for Mercedes Benz. Nana-Benz thus symbolises achievement and success as well as creativity and pride.

The economic liberalisation that swept most of the Global South in the 1980s and 1990s did not pass over West Africa. Rather, the lowering of tariff and non-tariff barriers became the order of the day. This resulted in new actors and new products entering the markets, including Chinese made 'West African' wax print. At first, the quality difference between the 'local' and the Chinese prints meant that the Nana-Benzes could keep a fair share of the profitable market and thus keep their position in society, but soon the entrance of Chinese textiles changed local power structures.

The changes were neither unilateral nor instant. Rather, a plethora of actors drove the change. They included Chinese traders trying to sell 'West African' wax prints on the local markets; Nana-Benzes who went to China to locate textile companies that could produce wax prints cheaper than the likes of Vlisco; and new young local actors, the so-called 'Les Nanettes', establishing close networks with Chinese producers and thus bypassing the dense network of the Nana-Benzes. The result was that Nana-Benzes lost their oligopolistic power and thus their ability to earn large profits from the trade. Instead, a larger group of Nanettes took over most of the sales; and products became much cheaper and sold in pieces of 3 metres rather than the original 6 – metre pieces – what some have referred to as the 'democratisation of the wax prints', allowing less wealthy traders to enter the trade and allowing ordinary citizens to buy the textiles.

Sources: (Sylvanus 2013; Axelsson and Sylvanus 2010; Prag 2013).

Political effects

There is no doubt that the revival of SSC – and in particular the revitalisation of China's interest in the rest of the Global South – has affected the political climate in the host economies. Crudely put, this revival has affected politics directly and indirectly: it has influenced populistic domestic politics especially during (Presidential) elections (direct effects) and it has affected some Southern economies' negotiating power vis-à-vis the Global North (indirect effects).

The most well-known example of how the Chinese came to overshadow a political campaign are the 2006 Presidential elections in Zambia. Sino-Zambian relations date back to the independence of Zambia in 1964 and even though the intensity of the relationship waxed and waned with domestic politics in China and with the geopolitical importance of Zambia for China, it was cast in mostly positive terms for 40 years. This changed overnight. In the final weeks of the 2006 – Presidential elections the then leader of one of the opposition parties, the Patriotic Front, Michael Chilufya Sata, criticised the presence of the Chinese and their investments in Zambia, calling them '*infestors*' – not investors, indicating that their economic activities had hidden motives. This message was received well by many Zambians living in urban areas. Positive notions such as collaboration, mutual benefits, and partnerships were replaced by accusations of flooding of cheap, low quality Chinese goods; use of Chinese convict labour; bad working conditions; large-scale repatriation of profits; tax exemptions; and squeezing the domestic private sector (Hampwaye and Kragelund 2013).

In the words of the then Vice-President of the Patriotic Front, Dr Guy Scott:

> *It's hard to know how they all got here . . . If you go to the market you find Chinese selling cabbages and beansprouts. What is the point in letting them in to do that? There's a lot of Chinese here doing construction. Zambians can do that. The Chinese building firms are undercutting the local firms . . . Our textile factories can't compete with cheap Chinese imports subsidised by a foreign government. People are saying: 'We've had bad people before. The whites were bad, the Indians were worse but the Chinese are worst of all.'*
>
> (cited in McGreal (2007)).

These anti-Chinese sentiments were partly linked to some fatal incidents in Chinese-owned companies, including the death of roughly 50 Zambians in an explosives factory, and partly to the populist platform from which the Patriotic Front campaigned during the 2006 Presidential elections (Larmer and Fraser 2007). Even though Michal Sata did not win the 2006 Presidential elections, the heated 'Zambia for Zambians' debate continued long after the elections and permeated subsequent policy initiatives. Most prominently, Sata's critique of the Chinese presence in Zambia became the cornerstone of Zambia's efforts to design indigenisation policies to bring ownership back to their own citizens (Kragelund 2012a).

When Sata eventually took over power in Zambia in 2011, his critique of the Chinese presence had evaporated. Most importantly, the global recession following the 2008 financial crisis had immediate impacts on the Zambian economy. Demand for primary commodities fell whereupon prices dropped immediately and Anglo-Saxon companies, in particular, had difficulties in refinancing debts and getting access to long-term finance for new operations. This led to huge layoffs, suspending contract labour, and reducing the scope of work for contracts and recruitment in the Zambian mining sector. In contrast, Chinese companies managed to keep production running at a high level and Chinese companies managed to purchase the closed mines. This coincided with a process of signing collective agreements with the unions leading to large increases in real wages for many employees of large-scale Chinese firms in Zambia, and the fact that the commodity boom – despite the 2008/9 setback – had driven economic growth in Zambia. In 2011, Chinese investors were thus saviours – not 'infestors'.

More indirectly, SSC has affected the policy space in the Global South, i.e. the ability of a state to define its own tailor-made developmental goals that reflect specific development challenges, and the availability of resources to attain these goals (UNCTAD 2007). This is closely linked to the country's ability to negotiate its aid deals with 'traditional' donors (Whitfield 2008). A number of cases seem to suggest that the availability of additional funds from Southern development partners have allowed countries in the Global South to – incrementally and episodically – enlarge their policy space either by playing out donors against each other or by attracting funds for projects not financed by 'traditional' donors.

Angola: turning to China after an IMF led aid embargo

In 2002, after decades of civil war, the Angolan government was ready to rebuild the country. The IMF, being the main gatekeeper for development finance to Angola, demanded both political and economic reforms in return for large-scale loans. The Angolan government, however, backed by large deposits of oil and rising commodity prices, refused to follow the neoliberal prescriptions laid down by the IFIs and announced that it would stop negotiations with the IMF, whereupon funds were cut. In their place, the Angolan state secured credit lines from the Chinese government to rebuild key infrastructure. At the beginning of 2011, credit lines worth approximately USD 14.5 billion had been provided from three Chinese state banks to Angola. This money funded some 100 infrastructure projects, all in line with the Angolan government's national development strategy. Meanwhile, the Angolan economy grew at an incredible two-digit speed and Angolan debt was rescheduled. The combination of vast oil resources, rapid economic growth during the decade since the end of the Angolan civil war, and the reduction of the debt burden has meant that several actors have made an effort to partner strategically with Angola. The improved situation has led Angola to secure credit lines worth more than USD 3 billion from countries as diverse as Germany, Spain, Canada, and Brazil (Corkin 2013; Mohan and Lampert 2012).

Cambodia: using development finance from the Southern partners to create 'balanced development'

Cambodia is among Asia's most aid dependant countries: roughly 90 per cent of public expenditures are financed by aid from bi-and multilateral donors. While 'traditional' donors like the USA, Germany, Australia, and the European Commission dominate aid for health and education, China, Japan, and South Korea lead aid to the transport sector in Cambodia (Sato et al. 2011). In fact, in 2009 'emerging' donors contributed 23.5 per cent of the total development assistance to the country (Greenhill, Prizzon, and Rogerson 2016: 143). Even if aid finances most of the public expenditures, the Cambodian government has for some time been dissatisfied with the 'traditional' donors as they only fund some of the projects described it its National Strategic Development Plan and the funding is perceived to be unpredictable. The transport sector, in particular, was underfunded by the 'traditional' donors relative to the stated needs of the country. Likewise, the Cambodian government has voiced its discontent with the lack of skills and knowledge transfer from the aid projects led by its 'traditional' partners.

While it is unclear whether or to what extent transfer of knowledge is any better from the 'emerging' donors, there is no doubt that 'emerging' donors help the Cambodian government achieve its infrastructure goals. In the first decade of the new millennium, China for instance transferred approximately USD 200 million in grants and another USD 500 million in concessional loans. The lion's share of this development finance targeted the transport and energy sectors including the refurbishment of national roads, extension of existing roads, construction of bridges, and feasibility studies on new railroads (Reilly 2012: Table 2).

SSC is thus seen as an option to finance the projects and sectors that are not prioritised by Northern development partners and the government is actively using the presence of 'emerging' donors to increase their 'negotiating capital' vis-à-vis the 'traditional' donors (Greenhill, Prizzon, and Rogerson 2016). In other words, the *Cambodian government accepts aid from emerging donors . . . as part of a carefully considered strategy that views the new donors as providing alternatives important to the country's balanced development*' (Sato et al. 2011: 2099).

Ethiopia: Using Chinese money to develop the energy sector

Recently, the Ethiopian government launched a Power Sector Master Plan to expand the capacity of the energy sector dramatically. Under normal circumstances the World Bank would finance hydropower projects like this, but in the end the expansion was financed by Chinese companies and Chinese development banks. Like the infrastructure for resources case in the Democratic Republic of Congo described in chapter 6, the main reason for China's involvement in Ethiopia is contrasting views on debt sustainability: whereas the IFIs see debt sustainability as prudent debt management, that is, the country's repayment ability via a revenue model, China perceives it as development sustainability, that is, the ability to repay the loans in the long term via economic development. Based on this difference, the World Bank was reluctant to finance the large hydraulic projects set in motion by the Ethiopian government. The very presence of China and its readiness to provide development finance for infrastructural projects, however, provided the Ethiopian government with policy space to develop its energy sector independently of the recommendations of the traditional partners (Feyissa 2012).

Nicaragua: welcoming partners that respect the national development agenda

During the past decade, Nicaragua has experienced major shifts in its donor relations. 'Northern' donors increasingly voice their concern over governance issues and poor human rights standards in the country leading them to phase out their aid to Nicaragua. Meanwhile, they have stopped providing budget support and instead, they either channel their funds to specific ministries (programme support) or bypass the Nicaraguan government altogether and instead work directly with civil society organisations. In contrast, 'Southern' donors have significantly increased their aid to the country.

The changing donor landscape results in the possibility of acquiring external finance for new types of projects for Nicaragua. In line with the Poverty Reduction Strategy Papers and the Millennium Development Goals, 'Northern' donors channelled most of their funds to social sectors in Nicaragua. 'Southern' donors, in contrast, have been willing to fund the productive investments that made up the central pillars of the National Human Development Plan (2012–2016), including a hydroelectric plant and a refinery.

Since Nicaragua joined the Bolivarian Alliance for the Peoples of Our America (ALBA) in 2007 (see chapter 3) Venezuela has been the most important Southern donor to Nicaragua. During the past decade, Venezuela has transferred more than USD 1.6 billion in aid, loans, and credit lines to Nicaragua. The majority of these funds have targeted the productive sector (and in particular the energy sector) but apparently funds are also allocated for agriculture and poverty reduction.

Moreover, the changing donor landscape has enabled the Nicaraguan government to welcome partners '*that are respectful of the government's vision for development . . . whereas those who want to manage their own agenda are not* [welcome]' (Walshe Roussel 2013: 808). While the wording is categorical the reality is more complex. In fact, it seems that the Nicaraguan government has managed to keep the money flowing from the 'Northern' donors while increasing the availability of funds via grants and loans from 'Southern' donors. Thereby, it has carved out policy space to fund its own development ideas (Walshe Roussel 2013).

Zambia: flexing muscles in relation to 'traditional' donors

In Zambia, a country long '*identified as an emblematic case of a country dominated by its donors*' (Fraser 2008b: 299), SSC has indeed been a contributing factor in explaining the creation of policy space. The mere visibility of China in the aid landscape has affected policy space in Zambia by allowing the government of Zambia to flex its muscles vis-à-vis the 'traditional' donors. It has resulted in the drafting of a new national development plan, without the direct involvement of the donors, and the recurrent critique of the DAC donors by key political figures in Zambia, asking them to 'pack their bags and go' if they keep interfering in internal affairs.

More importantly, however, booming commodity prices since the turn of the century have resulted in an ever larger share of the Zambian budget being made up of income from copper and cobalt and an ever smaller share coming from aid from 'traditional' partners. In fact, Zambia's dependence on aid has changed dramatically. In 2001, aid contributed some 53 per cent of the budget. Ten years later, in 2011,

Box 7.2 Chinese aid to the Malagasy health sector: popular but not sustainable

Supporting the health sector in the 'rest of the Global South' is high on the Chinese agenda. It is perceived as a way to improve political, economic, and cultural ties between China and the recipient country and thus facilitate larger foreign policy goals. China's health sector assistance to Madagascar is part of this greater scheme. It dates back to 1975 and is centred on Chinese medical teams that commit to two-year secondments in one of four Chinese health centres located in remote rural areas in the country.

The Chinese medical teams are composed of specialists such as gynaecologists, orthopaedic surgeons, pharmacists, and general medicine physicians, but they also always include acupuncturists and Chinese-medicine practitioners.

No doubt, the Chinese health centres fill a need in Madagascar. The services they offer are cheap compared to the national services, the medical teams are perceived as highly qualified, the acupuncturists and Chinese-medicine practitioners are well-received, and even if the Chinese pharmaceuticals offered in the centres were perceived by the customers to be of inferior quality compared to pharmaceuticals imported from the Global North this was counterbalanced by highly qualified staff.

Despite this, Kadetz and Hood (2017) conclude that China's support to the Malagasy health sector has failed. The reason is that despite discourses of mutual benefits and reciprocity, China has failed to build local capacity to run and maintain a well-functioning health system. In contrast, more than 40 years of health aid to Madagascar has left the country without skills and finances to run this. The Chinese team has not focused (sufficiently) on transferring skills and knowledge to local medical staff; technical instruction for equipment is only written in English; and the Chinese teams are bypassing the Malagasy Ministry of Health. In this sense, Chinese aid to the Malagasy health sector seems to resemble many aid projects originating in the Global North.

Source: (Kadetz and Hood 2017).

it contributed less than eight per cent. It was thus relatively easy for the Zambian government to flex its muscles vis-à-vis its 'traditional' partners (Kragelund 2014).

Social effects

One of the major controversies over the local effects of SSC relates to human rights (Taylor 2008). Not unlike the majority of SSC literature, the biggest controversies regarding social effects revolve around China's engagement in Africa. In short, the main argument is whether economic development will lead to social development and improvements of human rights (the Chinese approach) or whether development has to be rights-based from the onset (the 'Western' approach).

At the aggregate level the discussion has focused on whether or to what extent development finance from China, Venezuela, and oil-rich Middle Eastern states keep authoritarian regimes in power and thus undermine the possibility of 'good governance' reforms. Much anecdotal data has been provided to show that the Chinese principle of non-interference, its veto power in the United Nations Security Council, and its aid and credit lines keep dictators in power. However, cross-country data does not seem to support this claim. Based on data from 155 countries in the period from 1993 to 2008, Bader (2015: 23) concludes that '*China impacts autocratic survival much more weakly than critics of Beijing claim . . . Chinese arms sales, aid projects, and high-level diplomacy have no discernible impact on autocratic longevity*'.

At a more concrete level, the discussion has focused on employment conditions in Chinese and Indian firms in the 'Rest of the South', in particular. Abusive employment conditions like the use of child labour, extensive working hours, and poor health and safety standards have often been associated with Asian investments. These allegations are supported by Akorsu and Cooke (2011) who find that it is unlikely that Chinese and Indian multinational corporations will adopt high labour standards voluntarily. In the companies that they studied in Ghana, freedom of association was allowed but collective decisions were not respected. Moreover, they found relatively low wages and widespread discrimination based on gender and nationality in the companies.

The highly politicised nature of the Chinese presence in Zambia (see above) has attracted numerous researchers and civil society organisations. Among the latter, Human Rights Watch in 2011 commissioned a study on the labour practices in Chinese-owned mines in Zambia (Wells 2011). The result was a massive critique of the conditions in these companies. The study reports 12 – hours shifts in one department of a mine; 78 – hour workweeks in another; and work 365 – days a year in a third place. In a similar vein, it reports several accidents and health problems in the mines; mining corporations barring their employees from joining a union; and policies of active concealment of accidents.

Hairong and Sautman (2013) questions all of these conclusions. They argue that the interviewees in Human Rights Watch's study are not representative of the sector and that they most likely have been influenced by the anti-Chinese rhetoric of the Patriotic Front. Moreover, they argue that the study fails to compare fatality rates where Chinese mines come out on a par with other mining corporations, and that the study does not take recent improvements

---

Box 7.3 ProSAVANA: African agricultural powerhouse or large-scale land-grabbing?

In the spring of 2015, five years after the programme was first announced, the Ministry of Agriculture and Food Security, Mozambique, published the 'Zero Draft' of the master plan of the Program for Agricultural Development of the Tropical Savannah in Mozambique (ProSAVANA), a trilateral development cooperation between Japan, Brazil, and Mozambique built largely on experiences from a Japanese-Brazilian project in the central Brazilian *Cerrado* (savannah). The official aim of ProSAVANA is to improve and diversify agricultural production via public-private partnerships and thereby create opportunities for employment through the adoption of agricultural models developed and tested in Brazil. Thereby, the Ministry of Agriculture and Food Security perceives ProSAVANA as a means to transform Mozambique into an African agricultural powerhouse.

The Ministry of Agriculture and Food Security has not been able to persuade everybody that this is the only likely outcome of the trilateral development cooperation. In contrast, civil society organisations in both Brazil and Mozambique began contesting the project long before the 'Zero Draft' was officially released. They argue that rather than turning Mozambique into an African agricultural powerhouse, ProSAVANA will turn Mozambique into a haven for large-scale 'land-grabbing'. Their argument is that despite the obvious transformation of the Brazilian *Cerrado* – taming a 'wild west' and turning it into a centre of soya bean and sugarcane production and thus paving the way for Brazil's ethanol industry – the social and environmental costs have been massive. Most importantly, the transformation of the *Cerrado* has gone hand in hand with large-scale 'land-grabbing' by foreign investors.

In fact, the pre-release critique was so massive that the official 'Zero Draft' differs profoundly from the unofficial draft of ProSAVANA that was leaked in 2013. Most importantly, focus in the program changed from export-oriented crops to the inclusion of (local) smallholder farmers that focus on the domestic market. The Ministry of Agriculture and Food Security still wants to replicate the *Cerrado* via large-scale investments in export crops but this is now part of a larger programme to transform specific corridors – not the ProSAVANA.

Sources: (Clements and Fernandes 2013; Shankland and Gonçalves 2016).

into consideration. Likewise, the wage comparisons presented in the study are for unionised workers only thereby ignoring differences in workforce composition among mines. In contrast to this study, Hairong and Sautman (2013) conclude that labour practices indeed were problematic in the past but that massive improvements have taken place lately.

This methodological dispute is not ending the debate over the labour effects of SSC, however. In a recent study Isaksson and Kotsadam (2018b: 285) find that the presence of Chinese aid projects discourages trade union involvement in the area where the project is located and that this '*stems from direct anti-union policies*'. In short, they argue that the Chinese companies implementing the projects have no history of adhering to labour laws at home and hence, they induce pressure on local companies to not abide by labour laws either in order to be able to compete.

Conclusion

This chapter proposed an analytical framework to study the effects of SSC in the Global South. The framework distinguishes between direct and indirect impacts as well as between competitive and complementary. Thereby, it enables us to analyse both positive and negative effects of the rejuvenation of SSC and it informs us that the dichotomous presentations of for instance 'China in Africa' as being either good or bad for local development seldom tells the full story. No doubt, SSC can trigger economic benefits for the actors in recipient/host economies such as increase of financial flows to sectors and groups not targeted by the Global North; availability of cheap consumer goods (in smaller units) for poor consumers in both urban and rural areas; and availability of free education and skills development programmes. Simultaneously, this engagement can make recipients/hosts worse off: it may displace local production either through direct competition or because for instance China's exports to Europe displace Lesotho's export of the same commodities; and it may outmatch local entrepreneurs in both service, construction, and manufacturing sectors.

The chapter also demonstrated that it enhances our understanding to distinguish these economic effects from political and social effects. The political effects come in two different forms: either directly via influencing the political climate in the Global South or indirectly via affecting the policy space of recipient country governments enabling them to carve out better deals with other actors. However, it is important to bear in mind that most of these effects are short-lived and that they are often related to more fundamental political and economic changes, such as booming commodity prices, global economic recession, and global norms.

The social effects come in many forms. The most widely discussed form is that of rights. Southern actors like Venezuela, China, and United Arab Emirates have been accused of being rogue donors keeping dictators in power and benefitting from not abiding to basic labour rights. The picture on the ground, however, is less clear. Cross-country analysis cannot confirm that for instance Chinese development finance keeps authoritarian regimes in power. Likewise, data from host economies is mixed regarding the abuse of labour rights. What comes forth is that big Southern investors are now on average as good/bad as other investors but that a few years back there were more incidents of long working hours, low salaries, and prohibition on unionisation among Southern investors compared to Northern investors. It is worth mentioning, however, that we still know very little about how smaller privately-owned companies from China or India perform on the matters compared to smaller local firms or smaller European firms.

Discussion questions

- Explain the difference between the direct and indirect effects of SSC on the 'rest of the Global South' and provide examples of both competitive and complementary indirect impacts of this engagement.
- Describe the reasons why most studies of the economic effects of SSC end up in either end of a continuum – either praising the engagement highly or criticising it massively.
- Discuss the likelihood of China being a 'lead goose' making Africa the 'next factory of the world'.
- Consider what enlarged policy space means for social and economic development in the 'rest of the Global South'.

Web pages of interest

- Open University's Asian Drivers Programme was one of the first research programmes to systematically examine the effects of the rise of the Global South. The programme focused in particular on China and India and their effects on the rest of the Global South. Among the most important findings of the programme is the analytical framework presented in this chapter. The web site offers links to several studies of 'Asian giants' and their effects elsewhere. The programme ended in 2010 and hence some of the links are now defunct: asiandrivers.open.ac.uk/index.php

- Centre for Chinese Studies, Stellenbosch University offers policy relevant research on China and East Asia's impact on Africa: www.ccs.org.za
- The official web page for the ProSAVANA programme implemented by Brazil and Japan in Mozambique. It describes the sub-projects of the programme and links to important resources: www.prosavana.gov.mz

Further reading

Corkin, L. 2013. *Uncovering African Agency. Angola's Management of China's Credit Lines*. Farnham: Ashgate. This well-researched book sets out to correct the picture often depicted in the 'China-Africa' literature that China is always in the driver's seat. In contrast, Corkin shows that the political and econmic elite of Angola has managed the relationship with Chinese actors and used this relationship to bolster Angola's political capital vis-à-vis other actors interested in Angola's natural resources.

Sun, I. Y. 2017. *The Next Factory of the World: How Chinese Investment is Reshaping Africa*. Boston: Harvard Business Review. Sun's easily read book is based on visits to more than 50 Chinese-owned companies throughout Africa (clustered in four countries). Her main argument is that even if Chinese aid and infrastructure may be important for some African countries, the main contribution of China for Africa's future development is the Chinese migrants setting up factories throughout the continent. According to her data more than 10,000 Chinese-owend companies now exist in Africa and approximately one-third of these are in manufacturing. This latter group of companies will according to Sun bring about industrialisation in Africa.

Taylor, I. 2014. *Africa rising? BRICS-Diversifying dependency*. Suffolk: James Currey. In this well-written book Taylor challenges the much vaunted new world order created by the rise of the South where the rise of the 'emerging South' will lead to the rise of the 'rest of the Global South'. Instead, he argues that BRICS in Africa create structural dependency based on resource extraction resembling Africa's long-term engagement with the Global North.

8 Conclusion

According to UNOSSC (2018), SSC is '*a broad framework for collaboration among countries of the South in the political, economic, social, cultural, environmental, and technical domains ... Developing countries share knowledge, skills, expertise and resources to meet their development goals through concerted efforts*' and it continues: '*South-South cooperation is initiated, organized, and managed by developing countries themselves ...* [and] *The South-South cooperation agenda and South-South cooperation initiatives must be determined by the countries of the South, guided by the principles of respect for national sovereignty, national ownership and independence, equality, non-conditionality, non-interference in domestic affairs and mutual benefit*'.

This definition is not particularly operational. It neither informs the reader what SSC is nor how s/he recognises SSC when s/he sees it. More importantly, it only points to the intended positive outcomes of SSC whereas unintended, indirect, adverse effects are left out. This book has not aimed at providing a narrower definition of SSC, but it aims to cast open the black box of SSC, i.e. disaggregating the different vectors of engagement and the actors involved in driving current SSC development. Moreover, it has sought to illuminate the positive and negative, intended and unintended, direct and derived effects of SSC both on actors and on structures in the Global North as well as in the Global South. The account is by no means complete. Instead, this book has attempted to combine an account of the general ideas and main thrust of SSC with vignettes of how SSC play out in a variety of different locations and at different times. Moreover, it has made an effort to refer to analytical frameworks that the reader can apply to further his/her understanding of a particular variant of SSC. Ideally, the book therefore should provide the reader with a self-standing account of SSC and how it has developed over time and simultaneously point towards further reading. Below, I will draw up the main conclusions from the seven preceding chapters whereupon I will point to avenues of further research in the future.

Major SSC insights

This book established that what caught the eyes of journalists, politicians, and academics alike in Beijing in 2006 is not new at all. Rather, the 3rd FOCAC meeting and all the vectors of SSC engagement that we now experience build on more than 60 years of collaboration between countries, institutions, and peoples of the Global South. These actors make use of the same rhetoric and cast their collaboration in the same terms as they did during the Conference on Afro-Asian Peoples held in Bandung in 1955, but SSC today is not the same as SSC in the 1960s and 1970s. What really has changed is the (geo)political and economic context in which SSC plays out. Save for an institution like the Bolivarian Alliance for the Peoples of Our America – Peoples' Trade Agreement, none of the current SSC actors seek to radically change the overall system governing world affairs. In fact, it seems that the lion's share of actors engaging in SSC benefit from – and further promote – the current neoliberal world order. Even China's call for state-led development abides by the overall rules and norms of the world order. In contrast to the 1970s, SSC, therefore, depoliticises development and casts it in technical terms (Morvaridi and Hughes Forthcoming) and the actors seek to advance alternatives to the development hegemony of OECD-DAC within the overall framework.

It has also shown that Southern actors are heterogeneous and that competition exists between these actors. The heterogeneity stems not only from differences in size (Brazil vs. São Tomé and Príncipe) but also from their governance structures (India vs. China), their resource endowments (United Arab Emirates vs. Taiwan), their membership of global governance structures (China vs. Nigeria), and their historical role vis-à-vis the rest of the Global South (Cuba vs. South Africa). Therefore, it makes more sense to distinguish between two groups in the Global South, namely the 'Emerging South' (countries like Brazil, China, India, South Africa, and United Arab Emirates) and the 'rest of the Global South'. However, even within these two groups variety is massive. The 'rest of the Global South' includes both resource-rich and resource-poor countries; their levels of technological skills differ; their political systems differ; and the capacity of their state systems differ. Therefore, is comes as no surprise that SSC organisations are often at odds with each other. As depicted in chapter 3, the India, Brazil, South Africa Dialogue Forum (IBSA) is built on fundamentally different values compared to BRICS: while the former is made up of democratic countries, the latter is made up of economically, politically, and geographically very diverse nations. Moreover, the former came about as a reaction to the exclusory tendencies of the most powerful nations in the world whereas the latter was crafted as an investment idea and later taken up by very different, but increasingly powerful nations. Despite the non-uniform nature of BRICS compared to IBSA (and ALBA-TCP for that sake), it seems that it has outlived the other institutions due to its economic power derived predominantly from China.

Although most focus on SSC has been on aid – and especially Chinese aid – this book has revealed that SSC is about much more than aid but also that aid, or rather development finance, facilitates other much larger flows of exchange within the Global South. Even if data on these flows is still poor and definitions of, for instance, what makes up aid and how migrants are counted are still not comparable, there is no doubt that all vectors of South-South engagement are increasing. This intensification of course is not only a reflection of growing SSC but also mirrors the mounting economic power of the Global South in the world. Hence, if we are to understand the effects of the current phase of SSC we have to scrutinise all vectors of engagement and examine how they are mutually supporting each other. Trade, for example, has been facilitated by grand collaborative agreements that include pledges to lower tariff and non-tariff barriers on specific products. Likewise, investments have been facilitated by large-scale infrastructure for resources deals, and aid and investments have spurred an increase in migration that has stimulated further trade.

This being said, South-South aid has had an autonomous effect on the Global South both directly and indirectly. Directly, aid has been channelled to more sectors in more countries thereby hopefully filling a finance gap. Indirectly, South-South aid has introduced competitive pressures into the aid system. Now, donors from the Global North converge around concepts like 'win-win' and mutual benefit; they consequently focus more on the productive sectors – just like the donors of the Global South; and they merge forms of finance, i.e. a 'Southernisation' of development finance. Likewise, research on social welfare by Urbina-Ferretjans and

Surender (2013: 274) suggests '*that China's approach in the African continent is distinctly beginning to have an impact on Western ideas and activities*'. Whether the Global North also will focus less on civil society organisations, less on 'good governance', and more on projects aid in the future is still to be seen. However, it is not implausible as donor governments benefit more from the publicity of bilateral project aid to concrete projects than programme or budget aid where it is more difficult to trace the 'giver'.

The competitive pressure is not unidirectional: Southern donors are also changing their aid practices as a result of the engagement with other development actors. While South Korea, for instance, became a member of the DAC in 2010 and thereby transitioned from an 'emerging' to a 'traditional' donor, other Southern donors have taken active part in DAC's outreach activities and now share information about aid flows with DAC, partake in monitoring exercises, and contribute to discussions on the future direction of aid. Thereby, some of the different approaches to development even out over time. Even China, often portrayed as being poles apart from the DAC, is changing its aid practices incrementally. While its aid is still predominantly bilateral and project based, it now also focuses on many of the same topical issues and the same global public goods as the 'traditional donors'. No doubt, aid to productive sectors (and possibilities of mutual benefit) still dwarf aid to other sectors, but importantly here, China is increasing its focus on social sectors such as education and health. In the words of Urbina-Ferretjans and Surender (2013: 273), therefore, '*there are some signs of the possibility of China's norms changing as a consequence of greater interaction with Western institutions*'.

Despite the fact that we see these converging tendencies in the global aid regime, it is important to keep in mind that aid only constitutes an insignificant share of the total South-South financial flows. By far the most significant flows are trade, investment, and credit lines. Moreover, it is worth remembering that but for China and some OPEC-countries ODA-like flows are minuscule in absolute terms as well. We therefore also have to scrutinise trade and investments to get the full picture.

When we analyse South-South trade it is important to focus on changes in aggregate volumes and directions as well as changes in trade in parts as both types of changes inform us about the possibilities and barriers these transformations produce and for whom. This book has demonstrated that trade is definitely shifting towards the Global South but also that this growth is unevenly distributed across the Global South. While China and other East Asian manufacturing power houses account for an ever-bigger share of trade (and value added), countries in, for instance, sub-Saharan Africa have not experienced the same progress – neither in terms of volume nor in terms of value added. To understand this differentiated outcome we have to analyse how production is spatially distributed and where in this process value addition accrues. Analytically, the Global Value Chain approach is valuable. It describes input and output structures along a value chain, but more importantly for SSC, it points to how and to what extent producers in the Global South can benefit from participating in global trade. Moreover, it points to how governance of the chains affects upgrading possibilities for firms in the Global South. Thereby, it enables us to analyse whether linking up to a chain led by a firm in the Global South differs from linking up to a chain led by a firm in the Global North.

South-South investments have also been on the rise lately. Based on the scarce evidence available it seems that FDI from the Global South differs from FDI from the Global North in terms of sectoral focus, role of government in decision-making, and institutional environment. Whether this matters for developmental effects is an empirical question. It depends on the number of jobs created, whether skills and knowledge are transferred to local enterprises, the extent to which local companies are outcompeted, and the long-term social and environmental effects of the investment. According to one of the most comprehensive, though early, studies of characteristics of FDI from the Global South, Chinese and Indian companies tended to subcontract more from home country companies than comparable firms from the Global North thereby minimising the potential of positive spill overs (Broadman 2007).

Like trade and investments, South-South migration flows are also increasing. They are important as they on the one hand facilitate trade and investments and on the other hand may be a contributing factor in squeezing out local firms in the host societies.

These vectors of engagement do not work spontaneously. This book has established that they are driven – often intentionally – by a multitude of actors. The state in countries as diverse as South Africa, United Arab Emirates and China set up incentive schemes to facilitate the internationalisation of home country firms; they expand their representation abroad to smooth trade and investments, and they assist new migrants arriving in host societies. Moreover, they play an active role in changing the hearts and minds of the next generation of the political and economic elite in the 'rest of the Global South'. This is done via media and education programmes. It is worth noting though that the state often comprises several central and decentral entities and that these entities seldom follow a grand scheme. Rather, they work (semi)independently – all pursuing individual goals.

Moreover, state institutions rarely implement all the schemes set in motion as part of SSC. No doubt, state-owned enterprises do invest and government and provincial institutions do execute some aid projects, but the majority of SSC activities are implemented by private actors (and a few by civil actors). They win tenders and implement aid projects, they trade, and they invest throughout the Global South.

Furthermore, it is worth keeping in mind that trade and investments do not only flow in one direction: from the 'emerging South' to the 'rest of the Global South'. Rather, entrepreneurs and traders from for instance Africa benefit from the increased flows of money, goods, and ideas. As shown in this book, local manufacturers in the shoe and leather industry in Ethiopia are now exporting their goods to China. Likewise, African traders now go to China to buy goods specifically targeted at their home market. Often, they approach the producers directly making them change their designs to cater better for the new markets.

Finally, the book has presented an analytical framework to understand how SSC affects the Global South. The important aspect is to go beyond the direct effects and also examine indirect effects of SCC and also look for unintended negative effects of the collaboration. With this framework in mind it becomes clear that SSC influences the 'rest of the Global South' economically, politically, and socially both positively and negatively.

At the aggregate level, the outcome is neither solely positive nor solely negative economically: we are yet to see clear-cut examples of, for instance, China being a lead goose able to structurally transform African economies. Similarly, we do not only see examples of Chinese firms outmatching local ones, either. Rather, we see new economic avenues opening combined with stiffer competition in some places (and hence lower profit margins); we see new actors benefitting from new trade networks (and hence established ones losing out); and we see new (cheap) products entering the markets of the Global South competing with expensive ones from the Global North. Likewise, the political picture is not clear-cut either. Some aid recipient countries in the Global South have definitely benefitted from the competitive pressures introduced by Southern donors and thereby have managed to carve out a larger policy space that potentially can be used to negotiate better deals with 'traditional' donors. However, this effect is not only the result of more donors but also of rising commodity prices. Besides, this enlarged policy space has often been confined to specific sectors, it has been temporary and it has seldom changed the overall power structures between South and North (Carmody and Kragelund 2016; Taylor 2014). The social effects resemble the economic and political ones in terms of vagueness. The biggest issue seems to be related to rights; including labour rights. Evidence is still anecdotal but points towards differences between North and South being wiped out in terms of abuse of human rights. Instead, the new dividing line is between big visible companies (near major cities or along main transport corridors) with something at stake and smaller companies that can easily hide their malpractices from media and civil society organisations.

Future avenues of SSC research

The study of SSC has come a long way since the public reactions to the 3rd FOCAC reached the global media, but far too many studies still compare apples and oranges: ODA-like flows are compared to OOF-like flows, development finance is mixed with investments; and political statements are taken at face value. Moreover, data of poor quality is aggregated in large-scale cross-country comparisons to make bold statements about similarities or differences between vectors of engagement in the Global North and the Global South.

What we need now therefore is to qualify existing data. The AidData database described in chapter 1 is a case in point. It is a great step forward compared to the anecdotal evidence that existed before it was established but it still needs some improvements and refinements. Likewise, information on investments is far from accurate. The same goes for data on trade and migration. On top of this, the statistical data we use to measure changes in the Global South is far from perfect (Jerven 2014).

We also need in-depth studies of each separate vector of engagement and it is important that the booming 'China-Africa' literature is complemented by studies of 'China in Latin America' (Gallagher and Porzecanski 2010), by studies of the 'Middle East in Africa' and 'Africa in Africa'. Likewise, we need studies of spill overs from South-South FDI, of upgrading possibilities in value chains led by companies of the Global South, of the quality of construction work provided by Southern entrepreneurs, of enforcement of migration policies, and of the effects of short-and long-term education programmes offered by actors of the Global South. In a similar vein, it is of the utmost importance that we improve our knowledge of labour conditions and in particular, how Corporate Social Responsibility practices differ across different types of firms and how and to what extent home country characteristics matter for these practices. Studies like this will enable us to qualify the conclusions we put forward regarding the polycentric world that is currently emerging.

References

Abdenur, A. E., and J. M. E. M. Da Fonseca. 2013. The North's Growing Role in South – South Cooperation: keeping the foothold. *Third World Quarterly* 34 (8):1475–1491.

Adriansen, H. K., L. M. Madsen, and S. Jensen eds. 2015. *Higher Education and Capacity Building in Africa: The geography and power of knowledge under changing conditions.* New York: Routledge.

Akorsu, A. D., and F. L. Cooke. 2011. Labour standards application among Chinese and Indian firms in Ghana: typical or atypical? *International Journal of Human Resource Management* 22 (13):2730–2748.

Al-Mezaini, K. S. 2017. From Identities to Politics: UAE Foreign Aid. In *South-South Cooperation Beyond the Myths*, eds. I. Bergamaschi, P. Moore and A. B. Tickner, 225–244. New York: Palgrave Macmillan.

ALBA-TCP. 2017. *Bolivarian Alliance for the Peoples of Our America – Peoples' Trade Agreement (ALBA-TCP)*. ALBA-TCP 2017 [cited 9 October 2017].

Allen, K., and M. Galiano. 2017. Corporate Volunteering in the Global South. In *Perspectives on Volunteering: Voices from the South*, eds. J. Butcher and C. J. Einolf, 99–114. Cham: Springer.

Amighini, A., and M. Sanfilippo. 2014. Impact of South – South FDI and Trade on the Export Upgrading of African Economies. *World Development* 64 (Supplement C):1–17.

Apaydin, F. 2012. Overseas Development Aid across the Global South: Lessons from the Turkish Experience in Sub-Saharan Africa and Central Asia. *European Journal of Development Research* 24 (2):261–282.

Axelsson, L., and N. Sylvanus. 2010. Navigating Chinese textile networks: women traders in Accra and Lomé. In *The Rise of China & India in Africa*, eds. F. Cheru and C. Obi, 132–141. London: Zed Books.

Aykut, D., and R. Ratha. 2004. South-South FDI flows: how big are they? *Transnational Corporations* 13 (1):149–176.

Azmeh, S., and K. Nadvi. 2013. 'Greater Chinese' Global Production Networks in the Middle East: The Rise of the Jordanian Garment Industry. *Development and Change* 44 (6):1317–1340.

———— 2014. Asian firms and the restructuring of global value chains. *International Business Review* 23 (4):708–717.

Bader, J. 2015. China, Autocratic Patron? An Empirical Investigation of China as a Factor in Autocratic Survival. *International Studies Quarterly* 59 (1):23–33.

Bailey, M. 1975. Chinese aid in action: Building the Tanzania-Zambia railway. *World Development* 3 (7–8):587–593.

Baillie Smith, M., N. Laurie, and M. Griffiths. 2018. South – South volunteering and development. *The Geographical Journal* 184 (2):158–168.

Bakewell, O. 2009. South-south migration and human development: Reflections on African experiences. New York: UNDP.

Bandyopadhyaya, J. 1977. The Non-Aligned Movement and International Relations. *India Quarterly* 33 (2):137–164.

Benzi, D., and X. Zapata. 2017. Good-Bye Che?: Scope, Identity, and Change in Cuba's South – South Cooperation. In *South-South Cooperation Beyond the Myths*, eds. I. Bergamaschi, P. Moore and A. B. Tickner, 79–106. London: Palgrave Macmillan.

Biggeri, M., and M. Sanfilippo. 2009. Understanding China's move into Africa: an empirical analysis. *Journal of Chinese Economic and Business Studies* 7 (1):31–54.

Binder, A., and C. Meier. 2011. Opportunity knocks: Why non-Western donors enter humanitarianism and how to make the best of it. *International Review of the Red Cross* 93 (884):1135–1149.

Bodomo, A., and C. Pajancic. 2015. Counting Beans: Some Empirical and Methodological Problems for Calibrating the African Presence in Greater China. *Journal of Pan African Studies* 7 (10):126–143.

Bond, P. 2016. BRICS banking and the debate over sub-imperialism. *Third World Quarterly* 37 (4):611–629.

Bork-Hüffer, T., B. Rafflenbeul, Z. Li, F. Kraas, and D. Xue. 2016. Mobility and the Transiency of Social Spaces: African Merchant Entrepreneurs in China. *Population, Space and Place* 22 (2):199–211.

Bräutigam, D. 2009. *The Dragon's gift. The real story of China in Africa*. Oxford: Oxford University Press.

———— 2011. Aid 'with Chinese characteristics': Chinese foreign aid and development finance meet the OECD-DAC aid regime. *Journal of International Development* 23:752–764.

———— 2017. *Rubbery Numbers for Chinese Aid to Africa*. Johns Hopkins University, 30 April 2013 [cited 3 June 2017]. Available from www.chinaafricarealstory.com/2013/04/rubbery-numbers-on-chinese-aid.html

Bräutigam, D., and X. Tang. 2014. 'Going Global in Groups': Structural Transformation and China's Special Economic Zones Overseas. *World Development* 63:78–91.

Bräutigam, D., and T. Xiaoyang. 2011. African Shenzhen: China's special economic zones in Africa. *Journal of Modern African Studies* 49 (01):27–54.

Bräutigam, D., and J. Hwang. 2016. Eastern Promises: New Data on Chinese Loans in Africa, 2000 to 2014. Washington: School of Advanced International Studies, Johns Hopkins.

Bräutigam, D., T. Weis, and X. Tang. 2018. Latent advantage, complex challenges: Industrial policy and Chinese linkages in Ethiopia's leather sector. *China Economic Review* 48:158–169.

Braveboy-Wagner, J. A. 2016. Introduction: Rise of Which Global South States? In *Diplomatic Strategies of Nations in the Global South: The Search for Leadership*, ed. J. Braveboy-Wagner, 1–24. New York: Palgrave Macmillan US.

Breman, J. 2011. A Change for the Better? Dutch Development Aid in Good Times and Bad Times. *Development and Change* 42 (3):833–848.

Broadman, H. G. 2007. *Africa's Silk Road. China and India's New Economic Frontier*. Washington, D.C.: The World Bank.

Broich, T. 2017. Do authoritarian regimes receive more Chinese development finance than democratic ones? Empirical evidence for Africa. *China Economic Review* 46:180–207.

Brooks, A. 2010. Spinning and Weaving Discontent: Labour Relations and the Production of Meaning at Zambia-China Mulungushi Textiles. *Journal of Southern African Studies* 36 (1):113–132.

Carmody, P. 2009. Cruciform sovereignty, matrix governance and the scramble for Africa's oil: Insights from Chad and Sudan. *Political Geography* 28 (6):353–361.

——— 2011. *The new scramble for Africa*. Cambridge: Polity Press.

Carmody, P. 2013. *The rise of the BRICS in Africa. The geopolitics of South-South relations*. London: Zed Books.

Carmody, P., and P. Kragelund. 2016. Who is in Charge? State Power and Agency in Sino-African Relations. *Cornell International Law Journal* 49 (1):1–23.

Castillo, R. 2016. 'Homing' Guangzhou: Emplacement, belonging and precarity among Africans in China. *International Journal of Cultural Studies* 19 (3):287–306.

Chandra, V., J. Y. Lin, and Y. Wang. 2013. Leading Dragon phenomenon: new opportunities for catch-up in low-income countries. *Asian Development Review* 30 (1):52–84.

Chang, H.-J. 2003. The East Asian Development Experience. In *Rethinking development economics*, ed. H.-J. Chang, 107–124. London: Anthem Press.

Chant, S. H., and C. McIlwaine. 2009. *Geographies of development in the 21st century: an introduction to the global South*. Cheltenham: Edward Elgar Publishing.

Chapman, D. W., W. K. Cummings, and G. A. Postiglione. 2010. Transformations in Higher Education: Crossing Borders and Bridging Minds. In *Crossing Borders in East Asian Higher Education*, eds. D. W. Chapman, W. K. Cummings, and G. A. Postiglione, 1–22. Hong Kong: Springer.

Chaturvedi, S., T. Fues, and E. Sidiropoulos eds. 2012. *Development Cooperation and Emerging Powers*. London: Zed Books.

Cheru, F. 2016. Emerging Southern powers and new forms of South – South cooperation: Ethiopia's strategic engagement with China and India. *Third World Quarterly* 37 (4):592–610.

Cheru, F., and C. Obi. 2011. India-Africa relations in the 21st century: genuine partnership or a marriage of convenience? In *India in Africa. Changing Geographies of Power*, eds. E. Mawdsley and G. McCann. Cape Town: Pambazuka Press.

Chibuye, J., and S. Mvula. 2015. Zambia has 13,000 Chinese. *Zambia Daily Mail*, 21 March 2015.

China Daily. 2017. *Building mutual trust, brick by BRIC*. China Daily, 16 June 2009 [cited 9 October 2017]. Available from www.chinadaily.com.cn/china/2009–06/16/content_8286566.htm

Chun, H.-M., E. N. Munyi, and H. Lee. 2010. South Korea as an emerging donor: Challenges and changes on its entering OECD/DAC. *Journal of International Development* 22 (6):788–802.

Cisse, D. 2015. African traders in Yiwu: their trade networks and their role in the distribution of 'Made in China' products in Africa. *Journal of Pan African Studies* 7 (10):44–64.

Clements, E. A., and B. M. Fernandes. 2013. Land Grabbing, Agribusiness and the Peasantry in Brazil and Mozambique. *Agrarian South: Journal of Political Economy* 2 (1):41–69.

Cooper, A. F. 2017. The BRICS' New Development Bank: Shifting from Material Leverage to Innovative Capacity. *Global Policy* 8 (3):275–284.

Corea, G. 1977. UNCTAD and the new international economic order. *International Affairs* 53 (2):177–187.

Corkin, L. 2013. *Uncovering African Agency. Angola's Management of China's Credit Lines*. Farnham: Ashgate.

Cox, R. W. 1979. Ideologies and the New International Economic Order: reflections in some recent literature. *International Organization* 33 (2):257–302.

da Silva, A. L. R., A. P. Spohr, and I. L. da Silveira. 2016. From Bandung to Brasilia: IBSA and the political lineage of South – South cooperation. *South African Journal of International Affairs* 23 (2):167–184.

De Oliveira, R. S. 2008. Making sense of Chinese oil investments in Africa. In *China Returns to Africa. A Rising Power and a Continent Embrace*, eds. C. Alden, D. Large and R. Soares de Oliveira, 83–109. London: Hurst & Company.

——— 2015. *Magnificent and beggar land: Angola since the civil war*. Oxford: Oxford University Press.

Devereux, P. 2008. International volunteering for development and sustainability: outdated paternalism or a radical response to globalisation? *Development in Practice* 18 (3):357–370.

Dicken, P. 1998. *Global Shift. Transforming the World Economy*. 3rd ed. London: Paul Chapman Publishing.

Dirlik, A. 2007. Global South: predicament and promise. *The Global South* 1 (1):12–23.

Dobler, G. 2008a. From Scotch Whisky to Chinese Sneakers: International Comodity Flows and New Trade Networks in Oshikango, Namibia. *Africa* 78 (3):410–432.

——— 2008b. Solidarity, Xenophobia and the Regulation of Chinese businesses in Namibia. In *China Returns to Africa. A Rising Power and a Continent Embrace*, eds. C. Alden, D. Large and R. Soares de Oliveira, 237–256. London: Hurst & Company.

Donno, D., and N. Rudra. 2014. To Fear or Not to Fear? BRICs and the Developing World. *International Studies Review* 16 (3):447–452.

Dreher, A., P. Nunnenkamp, and R. Thiele. 2011. Are 'New' Donors Different? Comparing the Allocation of Bilateral Aid Between non-DAC and DAC Donor Countries. *World Development* 39 (11):1950–1968.

Dunning, J. H., C. Kim, and D. Park. 2008. Old wine in new bottles: a comparison of emerging market TNCs today and developed country TNCs thirty years ago. In *The Rise of Transnational Corporations from Emerging Markets: Threat or Opportunity?*, eds. K. P. Sauvant, K. Mendoza and I. Ince, 158–182. Cheltenham: Edward Elgar.

Eisenman, J., and J. Kurlantzick. 2006. China's Africa Strategy. *Current History* (May 2006):219–224.

Emerson, R. G. 2013. Institutionalising a Radical Region? The Bolivarian Alliance for the Peoples of Our America. *Journal of Iberian and Latin American Research* 19 (2):194–210.

Falk, P. S. 1987. Cuba in Africa. *Foreign Affairs* 65 (5):1077–1096.

Faria, C. A. P. d., and C. G. Paradis. 2013. Humanism and solidarity in Brazilian foreign policy under Lula (2003–2010): theory and practice. *Brazilian Political Science Review* 7: 8–36.

Farias, D. B. L. 2015. Triangular cooperation and the global governance of development assistance: Canada and Brazil as 'co-donors'. *Canadian Foreign Policy Journal* 21 (1): 1–14.

Farooki, M., and R. Kaplinsky. 2012. *The Impact of China on Global Commodity Prices. The global reshaping of the resource sector.* New York: Routledge.

Fejerskov, A. M. 2015. From Unconventional to Ordinary? The Bill and Melinda Gates Foundation and the Homogenizing Effects of International Development Cooperation. *Journal of International Development* 27 (7):1098–1112.

Fessehaie, J., and M. Morris. 2013. Value Chain Dynamics of Chinese Copper Mining in Zambia: Enclave or Linkage Development? *The European Journal of Development Research* 25 (4):537–556.

Feyissa, D. 2012. Aid negotiation: the uneasy 'partnership' between EPRDF and the donors. *Journal of Eastern African Studies* 5 (4):788–817.

Financial Tracking Service. 2018. *List of Government Donors.* UN Office for the Coordination of Humanitarian Affairs, ND 2018 [cited 2 July 2018]. Available from https://fts.unocha.org/donors/overview

Folke, S., N. Fold, and T. Enevoldsen. 1993. *South-South Trade and Development. Manufacturers in the New International Division of Labour.* New York: St. Martin's Press.

Fraser, A. 2008a. Aid-Recipient Sovereignty in Historical Context. In *The Politics of Aid. African Strategies for Dealing with Donors*, ed. L. Whitfield, 45–73. Oxford: Oxford University Press.

——— 2008b. Zambia: Back to the Future? In *The Politics of Aid. African Strategies for Dealing with Donors*, ed. L. Whitfield, 299–328. Oxford: Oxford University Press.

Fröbel, F., J. Heinrichs, and O. Kreye. 1978. The new international division of labour. *Information (International Social Science Council)* 17 (1):123–142.

G77. 2017. *About the Group of 77.* United Nations 2017 [cited 14 March 2017]. Available from www.g77.org/doc

Gallagher, K., and R. Porzecanski. 2010. *Dragon in the Room: China and the Future of Latin American Industrialization.* Stanford: Stanford University Press.

Gammeltoft, P. 2008. Emerging multinationals: Outward FDI from the BRICS countries. *International Journal of Technology and Globalisation* 4 (1):5–22.

Gammeltoft, P., H. Barnard, and A. Madhok. 2010. Emerging multinationals, emerging theory: Macro-and micro-level perspectives. *Journal of International Management* 16:95–101.

Gebre-Egziabher, T. 2007. Impacts of Chinese imports and coping strategies of local producers: the case of small-scale footwear enterprises in Ethiopia. *Journal of Modern African Studies* 45 (4):647–679.

——— 2009. The Developmental Impact of Asian Drivers on Ethiopia with Emphasis on Small-scale Footwear Producers. *World Economy* 32 (11):1613–1637.

Gebreeyesus, M., and P. Mohnen. 2013. Innovation Performance and Embeddedness in Networks: Evidence from the Ethiopian Footwear Cluster. *World Development* 41:302–316.

Gereffi, G. 1999. International trade and industrial upgrading in the apparel commodity chain. *Journal of International Economics* 48 (1):37–70.

Giannecchini, P., and I. Taylor. 2018. The eastern industrial zone in Ethiopia: Catalyst for development? *Geoforum* 88 (Supplement C):28–35.

Gibbon, P. 2003. The African Growth and Opportunity Act and the Global Commodity Chain for Clothing. *World Development* 31 (11):1809–1827.

Gibbon, P., and S. Ponte. 2005. *Trading Down. Africa, Value Chains, and the Global Economy*. Philadelphia: Temple University Press.

Gleijeses, P. 1997. The First Ambassadors: Cuba's Contribution to Guinea-Bissau's War of Independence. *Journal of Latin American Studies* 29 (1):45–88.

——— 2006. Moscow's Proxy? Cuba and Africa 1975–1988. *Journal of Cold War Studies* 8 (2):3–51.

Golub, P. S. 2013. From the New International Economic Order to the G20: how the 'global South' is restructuring world capitalism from within. *Third World Quarterly* 34 (6):1000–1015.

Government of China. 2006. China's Africa Policy.

Gray, K., and B. K. Gills. 2016. South – South cooperation and the rise of the Global South. *Third World Quarterly* 37 (4):554–557.

Greenhill, R., A. Prizzon, and A. Rogerson. 2016. The Age of Choice: Developing Countries in the New Aid Landscape. In *The Fragmentation of Aid: Concepts, Measurements and Implications for Development Cooperation*, eds. S. Klingebiel, T. Mahn and M. Negre, 137–151. London: Palgrave Macmillan UK.

Gu, J. 2009. China's Private Enterprises in Africa and the Implications for African Development. *European Journal of Development Research* 21 (4):570–587.

Gulrajani, N. 2016. Bilateral versus multilateral aid channels: strategic choices for donors. London: Overseas Development Institute.

Haglund, D. 2009. In It for the Long Term? Governance and Learning among Chinese Investors in Zambia's Copper Sector. *The China Quarterly* (199):627–646.

Hairong, Y., and B. Sautman. 2012. Chasing Ghosts: Rumours and Representations of the Export of Chinese Convict Labour to Developing Countries. *The China Quarterly* 210:398–418.

——— 2013. 'The Beginning of a World Empire'? Contesting the Discourse of Chinese Copper Mining in Zambia. *Modern China* 39 (2):131–164.

Hampwaye, G., and P. Kragelund. 2013. Trends in Sino-Zambian Relations. In *China's Diplomacy in Eastern and Southern Africa*, ed. S. Adem, 27–40. Farnham: Ashgate.

Harlan, T. 2017. A green development model: transnational model-making in China's small hydropower training programmes. *Area Development and Policy* 2 (3):251–271.

Harman, S., and W. Brown. 2013. In from the margins? The changing place of Africa in International Relations. *International Affairs* 89 (1):69–87.

Hart, G. 2002. *Disabling Globalization. Places of Power in Post-Apartheid South Africa*. Berkeley: University of California Press.

Haugen, H. Ø. 2013. China's recruitment of African university students: policy efficacy and unintended outcomes. *Globalisation, Societies and Education* 11 (3):315–334.

Haugen, H. Ø. 2018. Petty commodities, serious business: the governance of fashion jewellery chains between China and Ghana. *Global Networks* 18 (2):307–325.

Haugen, H. Ø., and J. Carling. 2005. On the edge of the Chinese diaspora: The surge of baihuo business in an African city. *Ethnic and Racial Studies* 28 (4):639–662.

Henry, M. 2012. Peacexploitation? Interrogating Labor Hierarchies and Global Sisterhood Among Indian and Uruguayan Female Peacekeepers. *Globalizations* 9 (1):15–33.

Hernandez, D. 2017. Are 'New' Donors Challenging World Bank Conditionality? *World Development* 96:529–549.

Hofmeyr, I., and D. Betty Govinden. 2008. Africa/India: Culture and circulation in the Indian Ocean. *Scrutiny 2* 13 (2):5–15.

Horner, R. 2016. A New Economic Geography of Trade and Development? Governing South – South Trade, Value Chains and Production Networks. *Territory, Politics, Governance* 4 (4):400–420.

Horner, R., and K. Nadvi. Forthcoming. Global value chains and the rise of the Global South: unpacking twenty-first century polycentric trade. *Global Networks*:n/a-n/a.

Hunt, D. 1989. *Economic theories of development: an analysis of competing paradigms*. Hemel Hempstead: Harvester Wheatsheaf.

Hynes, W., and S. Scott. 2013. *The Evolution of Official Development Assistance*. Paris: OECD Publishing.

Isaksson, A.-S., and A. Kotsadam. 2018a. Chinese aid and local corruption. *Journal of Public Economics* 159:146–159.

———— 2018b. Racing to the bottom? Chinese development projects and trade union involvement in Africa. *World Development* 106:284–298.

Jansson, J. 2013. The Sicomines agreement revisited: prudent Chinese banks and risk-taking Chinese companies. *Review of African Political Economy* 40 (135):152–162.

Jerven, M. 2014. *Economic Growth and Measurement Reconsidered in Botswana, Kenya, Tanzania, and Zambia, 1965–1995*. Oxford: Oxford University Press.

Kadetz, P., and J. Hood. 2017. Outsourcing China's welfare: unpacking the outcomes of 'sustainable' self-development in Sino-African health diplomacy. In *Handbook of Welfare in China*, eds. B. Carillo, J. Hood and P. Kadetz, 338–360. Cheltenham: Edward Elgar.

Kaplinsky, R., and D. Messner. 2008. Introduction: The Impact of Asian Drivers on the Developing World. *World Development* 36 (2):197–209.

Karshenas, M. 2016. Power, Ideology and Global Development: On the Origins, Evolution and Achievements of UNCTAD. *Development and Change* 47 (4):664–685.

Keijzer, N., and E. Lundsgaarde. 2017. When unintended effects become intended: implications of 'mutual benefit' discourses for development studies and evaluation practices. Nijmegen: Ministry of Foreign Affairs of the Netherlands and Radboud University.

Kim, E. M., and J. E. Lee. 2013. Busan and beyond: South Korea and the transition from aid effectiveness to development effectiveness. *Journal of International Development* 25 (6):787–801.

King, K. 2013. *China's Aid and Soft Power in Africa*. Woodbridge: James Currey.

Klare, M., and D. Volman. 2006. America, China & the Scramble for Africa's Oil. *Review of African Political Economy* 33 (108):297–309.

Kojima, K. 2000. The 'flying geese' model of Asian economic development: origin, theoretical extensions, and regional policy implications. *Journal of Asian Economics* 11 (4):375–401.

Konings, P. 2007. China and Africa in the Era of Neo-Liberal Globalisation. *Codesria Bulletin* (1&2):17–22.

Kot-Majewska, K. 2015. Role of Non-traditional Donors in Humanitarian Action: How Much Can They Achieve? In *The Humanitarian Challenge: 20 Years European Network on Humanitarian Action (NOHA)*, eds. P. Gibbons and H.-J. Heintze, 121–134. Cham: Springer International Publishing.

Kragelund, P. 2008. The return of Non-DAC donors to Africa: New prospects for African Development? *Development Policy Review* 26 (5):555–584.

————— 2009a. Knocking on a wide open door: Chinese investments in Africa. *Review of African Political Economy* 36 (122):479–497.

————— 2009b. Part of the Disease or Part of the Cure? Chinese Investments in the Zambian Mining and Construction sectors. *European Journal of Development Research* 21 (4):644–661.

————— 2011. Back to BASICs? The Rejuvenation of Non-traditional Donors' Development Cooperation with Africa. *Development and Change* 42 (2):585–607.

————— 2012a. Bringing 'indigenous' ownership back: Chinese presence and the Citizen Economic Empowerment Commission in Zambia. *Journal of Modern African Studies* 50 (3):447–466.

————— 2012b. The Revival of Non-Traditional State Actors' Interests in Africa: Does it Matter for Policy Autonomy? *Development Policy Review* 30 (6):703–718.

————— 2014. 'Donors go home': non-traditional state actors and the creation of development space in Zambia. *Third World Quarterly* 35 (1):145–162.

————— 2015. Towards convergence and cooperation in the global development finance regime: closing Africa's policy space? *Cambridge Review of International Affairs* 28 (2):246–262.

Kragelund, P., and P. Carmody. 2016. The BRICS' impacts on local economic development in the Global South: the cases of a tourism town and two mining provinces in Zambia. *Area Development and Policy* 1 (2):218–237.

Kragelund, P., and G. Hampwaye. 2016. The Confucius Institute at the University of Zambia: a new direction in the internationalisation of African higher education? In *Higher Education and Capacity Building in Africa*, 83–104. London: Routledge.

Kristinsson, T. 2017. Rising Powers and the End of Colonial Trade Patterns. In *EADI Nordic Conference 2017: Globalisation at the Crossroads – Rethinking Inequalities and Boundaries*. Bergen, Norway.

Lancaster, C. 2008. *Foreign aid: Diplomacy, development, domestic politics*. Chicago: University of Chicago Press.

Large, D. 2008. Beyond 'Dragon in the Bush': the Study of China-Africa Relations. *African Affairs* 107 (426):45–61.

Large, D., and L. A. Patey. 2011. China, India & the Politics of Sudan's Asian Alternatives. In *Sudan Looks East: China, India & the Politics of Asian Alternatives*, eds. D. Large and L. A. Patey, 176–194. Woodbridge: James Currey.

Larmer, M., and A. Fraser. 2007. Of cabbages and King Cobra: Populist politics and Zambia's 2006 election. *African Affairs* 106 (425):611–637.

Lauridsen, L. S. 2008. *State, institutions and industrial development: industrial deepening and upgrading policies in Taiwan and Thailand compared*. Aachen: Shaker Verlag.

Lee, C. K. 2009. Raw Encounters: Chinese Managers, African Workers and the Politics of Casualization in Africa's Chinese Enclaves. *The China Quarterly* (199):647–666.

Leftwich, A. 1995. Bringing politics back in: Towards a model of the developmental state. *Journal of Development Studies* 31 (3):400–427.

Li, A. 2007. China and Africa: Policy and Challenges. *China Security* 3 (3):69–93.

Lim, G. 2017. What Do Malaysian Firms Seek in Vietnam? *Journal of Asia-Pacific Business* 18 (2):131–150.

Lin, J. Y. 2012. From Flying Geese to Leading Dragons: New Opportunities and Strategies for Structural Transformation in Developing Countries. *Global Policy* 3 (4):397–409.

Linares, R. 2011. The ALBA alliance and the construction of a new Latin American regionalism. *International Journal of Cuban Studies* 2 (1/2):145–156.

Lututala, B. M. 2014. Intra-and Extraregional Migration in the South: The Case of Africa. In *A New Perspective on Human Mobility in the South*, 21–47. Dordrecht: Springer.

Lyons, M., A. Brown, and L. Zhigang. 2012. In the Dragon's Den: African Traders in Guangzhou. *Journal of Ethnic and Migration Studies* 38 (5):869–888.

Malm, J. 2016. When Chinese development finance met the IMF's public debt norm in DR Congo. PhD thesis, Department of Social Sciences and Business, Roskilde Universitet, Roskilde.

Manning, R. 2006. Will 'Emerging Donors' Change the Face of International Co-operation? *Development Policy Review* 24 (4):371–385.

Masters, L. 2014. Building bridges? South African foreign policy and trilateral development cooperation. *South African Journal of International Affairs* 21 (2):177–191.

Mawdsley, E. 2008. Fu Manchu versus Dr Livingstone in the Dark Continent? Representing China, Africa and the West in British broadsheet newspapers. *Political Geography* 27:509–529.

———— 2010. Non-DAC donors and the changing landscape of foreign aid: the (in)significance of India's development cooperation with Kenya. *Journal of Eastern African Studies* 4 (2):361–379.

———— 2011. The Conservatives, the Coalition and international development. *Area* 43 (4):506–507.

———— 2012a. The changing geographies of foreign aid and development cooperation: contributions from gift theory. *Transactions of the Institute of British Geographers* 37 (2):256–272.

———— 2012b. *From Recipients to Donors. Emerging Powers and the Changing Development Landscape*. London: Zed Books.

———— 2015. DFID, the Private Sector and the Re-centring of an Economic Growth Agenda in International Development. *Global Society* 29 (3):339–358.

———— 2017a. Development geography 1: Cooperation, competition and convergence between 'North' and 'South'. *Progress in Human Geography* 41 (1):108–117.

———— 2017b. National interests and the paradox of foreign aid under austerity: Conservative governments and the domestic politics of international development since 2010. *The Geographical Journal* 183 (3):223–232.

Mawdsley, E., and G. McCann eds. 2011. *India in Africa: Changing Geographies of Power*. Cape Town: Pambazuka Press.

Mawdsley, E., L. Savage, and S.-M. Kim. 2014. A 'post-aid world'? Paradigm shift in foreign aid and development cooperation at the 2011 Busan High Level Forum. *The Geographical Journal* 180 (1):27–38.

McCormick, D., P. Kamau, and P. Ligulu. 2006. Post-Multifibre Arrangement Analysis of the Textile and Garment Sectors in Kenya. *IDS Bulletin* 37 (1):80–88.

McEwan, C., and E. Mawdsley. 2012. Trilateral Development Cooperation: Power and Politics in Emerging Aid Relationships. *Development and Change* 43 (6):1185–1209.

McGreal, C. 2010. *Thanks China, now go home: buy-up of Zambia revives old colonial fears*. The Guardian, February 7 2007 [cited 12 August 2010]. Available from www.guardian.co.uk/world/2007/feb/05/china.chrismcgreal

Menzel, U. 1983. Der Differenzierungsprozeß in der Dritten Welt und seine Konsequenzen für den Nord-Süd-Konflikt und die Entwicklungstheorie. *Politische Vierteljahresschrift* 24 (1):31–59.

Milhorance, C., and F. Soule-Kohndou. 2017. South-South Cooperation and Change in International Organizations. *Global Governance* 23 (3):461–481.

Mohan, G., and B. Lampert. 2012. Negotiating China: Reinserting African Agency into China-Africa Relations. *African Affairs* 112 (446):92–110.

Mohan, G., and M. Tan-Mullins. 2009. Chinese Migrants in Africa as New Agents of Development? An Analytical Framework. *European Journal of Development Research* 21 (4):588–605.

Momani, B., and C. A. Ennis. 2012. Between caution and controversy: lessons from the Gulf Arab states as (re-)emerging donors. *Cambridge Review of International Affairs* 25 (4):605–627.

Monson, J. 2010. *Africa's Freedom Railway: How a Chinese Development Project Changed the Lives and Livelihoods in Tanzania.* Indiana University Press: Bloomington.

Morphet, S. 2004. Multilateralism and the Non-Aligned Movement: What Is the Global South Doing and Where Is It Going? *Global Governance* 10 (4):517–537.

Morvaridi, B., and C. Hughes. Forthcoming. South – South Cooperation and Neoliberal Hegemony in a Post-aid World. *Development and Change* 49 (3):867–892.

Muchapondwa, E., D. Nielson, B. Parks, A. M. Strange, and M. J. Tierney. 2016. 'Ground-Truthing' Chinese Development Finance in Africa: Field Evidence from South Africa and Uganda. *The Journal of Development Studies* 52 (6):780–796.

Muhr, T. 2011. Conceptualising the ALBA-TCP: third generation regionalism and political economy. *International Journal of Cuban Studies*:98–115.

——— 2016. Beyond 'BRICS': ten theses on South – South cooperation in the twenty-first century. *Third World Quarterly* 37 (4):630–648.

Mwase, N. 1983. The Tanzania-Zambia Railway: The Chinese Loan and the Pre-Investment Analysis Revisited. *The Journal of Modern African Studies* 21 (3):535–543.

Naidoo, R. 2011. Rethinking development: Higher education and the new imperialism. In *Handbook on globalization and higher education*, eds. R. King, S. Marginson and R. Naidoo, 40–58. Cheltenham: Edward Elgar.

Naidu, S., and D. Mbazima. 2008. China-African relations: A new impulse in a changing continental landscape. *Futures* 40 (8):748–761.

Naím, M. 2007. Rogue aid. *Foreign Policy* March/April 2007 (159).

Ndjio, B. 2009. 'Shanghai Beauties' and African Desires: Migration, Trade and Chinese Prostitution in Cameroon. *The European Journal of Development Research* 21 (4):606–621.

——— 2017. Sex and the transnational city: Chinese sex workers in the West African city of Douala. *Urban Studies* 54 (4):999–1015.

Newman, C., J. Page, J. Rand, and A. Shimeles. 2016. *Made in Africa: Learning to compete in industry.* Washington DC: Brookings Institution Press.

O'Hagan, J., and M. Hirono. 2014. Fragmentation of the International Humanitarian Order? Understanding 'Cultures of Humanitarianism' in East Asia. *Ethics & International Affairs* 28 (4):409–424.

O'Neill, J. 2001. Building better global economic BRICs. In *Global Economics Paper*. Newark: Goldman Sachs,.

OECD. 2010. *Perspectives on Global Development 2010. Shifting Wealth.* Paris: OECD Development Centre.

——— 2017. *Becoming a Participant in the Development Assistance Committee (DAC).* OECD, ND ND-a [cited 19 October 2017]. Available from www.oecd.org/dac/dac-global-relations/Becoming_a_Participant_in_the_DAC.pdf

——— 2017. *United Arab Emirates' Development Co-operation.* OECD, ND ND-b [cited 11 November 2017]. Available from www.oecd.org/dac/stats/uae-official-development-assistance.htm

Özkan, M., and B. Akgün. 2010. Turkey's opening to Africa. *The Journal of Modern African Studies* 48 (4):525–546.

Pacitto, J., and E. Fiddian-Qasmiyeh. 2013. Writing the 'Other' into humanitarian discourse: framing theory and practice in South – South humanitarian responses to forced displacement. Oxford: Refugee Studies Centre, Oxford Department of International Development, University of Oxford.

Pan, L. ed. 1999. *The Encyclopedia of the Chinese Overseas.* Richmond: Curzon Press.

Park, Y. J. 2009. *Chinese migration in Africa*: Johannesburg: South African Institute of International Affairs.

Patey, L. A. 2009. Against the Asian tide: the Sudan divestment campaign. *Journal of Modern African Studies* 47 (4):551–573.

——— 2014. *The New Kings of Crude: China, India, and the Global Struggle for Oil in Sudan and South Sudan.* London: Hurst.

——— 2017. Learning in Africa: China's Overseas Oil Investments in Sudan and South Sudan. *Journal of Contemporary China* 26 (107):756–768.

Paulo, S., and H. Reisen. 2010. Eastern Donors and Western Soft Law: Towards a DAC Donor Peer Review of China and India? *Development Policy Review* 28 (5):535–552.

Pedersen, M. A., and M. Nielsen. 2013. Trans-temporal hinges: reflections on an ethnographic study of Chinese infrastructural projects in Mozambique and Mongolia. *Social Analysis* 57 (1):122–142.

Phelps, N. A., J. C. H. Stillwell, and R. Wanjiru. 2009. Broken Chain? AGOA and Foreign Direct Investment in the Kenyan Clothing Industry. *World Development* 37 (2):314–325.

Pomeroy, M., A. Shankland, A. Poskitt, K. K. Bandyopadhyay, and R. Tandon. 2016. Civil Society, BRICS and International Development Cooperation: Perspectives from India, South Africa and Brazil. In *The BRICS in International Development*, eds. J. Gu, A. Shankland and A. Chenoy, 169–206. London: Palgrave Macmillan UK.

Postel, H. 2017. *Following the Money: Chinese Labor Migration to Zambia* [Feature]. Migration Policy Institute, February 20 2015 [cited 18 October 2017]. Available from www.migrationpolicy.org/article/following-money-chinese-labor-migration-zambia

——— 2017. *We may have been massively overestimating the number of Chinese migrants in Africa* [Blog]. African Arguments, 19 December 2016 [cited 17 October 2017]. Available from www.africanarguments.org/2016/12/19/we-may-have-been-massively-overestimating-the-number-of-chinese-migrants-in-africa

Prag, E. 2013. Mama Benz in Trouble: Networks, the State, and Fashion Wars in the Beninese Textile Market. *African Studies Review* 56 (3):101–121.

Pruitt, L. J. 2016. *The women in blue helmets: gender, policing, and the UN's first all-female peacekeeping unit.* Oakland: University of California Press.

Quadir, F. 2013. Rising Donors and the New Narrative of 'South – South' Cooperation: what prospects for changing the landscape of development assistance programmes? *Third World Quarterly* 34 (2):321–338.

Ramamurti, R. 2009. Why study emerging-market multinationals? In *Emerging Multinationals in Emerging Markets*, eds. R. Ramamurti and J. V. Singh, 3–22. Cambridge: Cambridge University Press.

Ratha, D., C. Eigen-Zucchi, and S. Plaza. 2016. *Migration and remittances Factbook 2016*. Washington DC: World Bank Publications.

Ratha, D., and W. Shaw. 2007. *South-South migration and remittances*. Washington DC: World Bank Publications.

Reilly, J. 2012. A Norm-Taker or a Norm-Maker? Chinese aid in Southeast Asia. *Journal of Contemporary China* 21 (73):71–91.

Richey, L. A., and S. Ponte. 2014. New actors and alliances in development. *Third World Quarterly* 35 (1):1–21.

Riddell, R. C. 2007. *Does Foreign Aid Really Work?* Oxford: Oxford University Press.

Rigg, J. 2007. *An everyday geography of the global south*. New York: Routledge.

Rist, G. 2008. *The history of development: from western origins to global faith (3rd edition)*. London: Zed Books.

Roca, S. 1980. Economic aspects of Cuban involvement in Africa/Aspectos econbmicos de la presencia Cubana en Africa. *Cuban Studies* 10 (2):55–80.

Sato, J., H. Shiga, T. Kobayashi, and H. Kondoh. 2011. 'Emerging Donors' from a Recipient Perspective: An Institutional Analysis of Foreign Aid in Cambodia. *World Development* 39 (12):2091–2104.

Selvaratnam, V. 1988. Higher education co-operation and Western dominance of knowledge creation and flows in Third World countries. *Higher Education* 17 (1):41–68.

Shankland, A., and E. Gonçalves. 2016. Imagining Agricultural Development in South – South Cooperation: The Contestation and Transformation of ProSAVANA. *World Development* 81:35–46.

Shushan, D., and C. Marcoux. 2011. The Rise (and Decline?) of Arab Aid: Generosity and Allocation in the Oil Era. *World Development* 39 (11):1969–1980.

Söderbaum, F. 2004. *The political economy of regionalism: the case of Southern Africa*. Houndmills: Palgrave Macmillan.

Strauss, J. C. 2013. China and Africa Rebooted: Globalization(s), Simplification(s), And Cross-cutting Dynamics in 'South-South' Relations. *African Studies Review* 56 (1):155–170.

Stuenkel, O. 2015a. *The BRICS and the future of global order*. London: Lexington Books.

———— 2015b. *India-Brazil-South Africa dialogue forum (IBSA): the rise of the global south?* London: Routledge.

Sumner, A. 2012. Where Do The Poor Live? *World Development* 40 (5):865–877.

Sun, I. Y. 2017. *The Next Factory of the World: How Chinese Investment is Reshaping Africa*. Boston: Harvard Business Review.

Sylvanus, N. 2013. Chinese devils, the global market, and the declining power of Togo's Nana-Benzes. *African Studies Review* 56 (1):65–80.

Tan, G. 1993. The next NICS of Asia. *Third World Quarterly* 14 (1):57–73.

Taylor, I. 2006. China's oil diplomacy in Africa. *International Affairs* 82 (5):937–959.

———— 2008. Sino-African Relations and the Problem of Human Rights. *African Affairs* 107 (426):63–87.

———— 2011. *The Forum on China-Africa Cooperation (FOCAC)*. New York: Routledge.

———— 2012. India's rise in Africa. *International Affairs* 88 (4):779–798.

———— 2014. *Africa rising? BRICS-Diversifying dependency*. Suffolk: James Currey.

The Economist. 2017. *Hopeless Africa*. The Economist, 11 May 2000 [cited 9 October 2017]. Available from www.economist.com/node/333429

———— 2006. China in Africa. Never too late to scramble. *The Economist*, 28 October 2006.

——— 2017. *The hopeful continent. Africa Rising*. The Economist, 3 December 2011 [cited 8 February 2017]. Available from www.economist.com/node/21541015

Thorbecke, E. 2007. The evolution of the development doctrine, 1950–2005. In *Advancing Development: Core Themes in Global Economics*, eds. G. Mavrotas and A. Shorrocks, 3–36. Houndmills: Palgrave Macmillan.

Tomlinson, B. R. 2003. What Was the Third World? *Journal of Contemporary History* 38 (2):307–321.

Tull, D. M. 2006. China's engagement in Africa: scope, significance and consequences. *Journal of Modern African Studies* 44 (3):459–479.

UN Volunteers. 2018. *Words of welcome*. United Nations Development Programme, ND 2018 [cited 2 July 2018]. Available from www.unv.org/annual-report/annual-report-2017

UNCTAD. 2005. South-South Cooperation in International Investment Arrangements. In *UNCTAD Series on International Investment Policies for Development*, 96. New York: United Nations Conference on Trade and Development.

——— 2006. *World Investment Report 2006. FDI from Developing and Transition Economies: Implications for Development*. New York and Geneva: UNCTAD.

——— 2007. Reclaiming Policy Space. Domestic Resource Mobilization and Developmental States. In *Economic Development Report in Africa 2007*. New York and Geneva: UNCTAD.

——— 2010. South-South Cooperation: Africa and the New Forms of Development Partnership. New York and Geneva: UNCTAD.

——— 2015. Global Value Chains and South-South Trade. Economic Cooperation and Integration among Developing Countries. New York and Geneva: UNCTAD.

UNDP. 1974. Report on The Working Group on Technical Co-operation among Developing Countries on its Third Session. Washington DC: United Nations Development Programme.

——— 1994. The Buenos Aires Plan of Action for Promoting and Implementing Technical Co-operation among Developing Countries. New York: United Nations Development Programme.

——— 2017. *IBSA Fund*. United Nations Office for South-South Cooperation, ND 2017 [cited 16 October 2017]. Available from http://tcdc2.undp.org/ibsa

——— 2018. *Human Development Reports*. United Nations Development Programme, ND 2018 [cited 18 June 2018]. Available from http://hdr.undp.org/en/composite/HDI

United Nations. 2009a. Promotion of South-South cooperation for development: a thirty-year perspective. Report of the Secretary General. New York: United Nations General Assembly.

——— 2009b. The state of South-South Cooperation. Report of the Secretary General. New York: United Nations General Assembly.

UNOSSC. 2018. *What Is South-South Cooperation?* United Nations Office for South-South Cooperation 2018 [cited 20 March 2018]. Available from http://unossc1.undp.org/sscexpo/content/ssc/about/what_is_ssc.htm

Urbina-Ferretjans, M., and R. Surender. 2013. Social policy in the context of new global actors: How far is China's developmental model in Africa impacting traditional donors? *Global Social Policy* 13 (3):261–279.

van der Merwe, J. 2016. Seeing Through the MIST: New Contenders for the African Space? In *Emerging Powers in Africa: A New Wave in the Relationship?*, eds. J. van der Merwe, I. Taylor and A. Arkhangelskaya, 1–14: Springer.

Verschaeve, J., and J. Orbie. 2016. The DAC is Dead, Long Live the DCF? A Comparative Analysis of the OECD Development Assistance Committee and the UN Development Cooperation Forum. *European Journal of Development Research* 28 (4):571–587.

Vestergaard, J., and R. H. Wade. 2013. Protecting Power: How Western States Retain The Dominant Voice in The World Bank's Governance. *World Development* 46:153–164.

——— 2015. Still in the Woods: Gridlock in the IMF and the World Bank Puts Multilateralism at Risk. *Global Policy* 6 (1):1–12.

Vieira, M. A. 2013. IBSA at 10: South – South development assistance and the challenge to build international legitimacy in a changing global order. *Strategic Analysis* 37 (3):291–298.

Villanger, E. 2007. Arab Foreign Aid: Disbursement Patterns, Aid Policies and Motives. *Forum for Development Studies* 34 (2):223–256.

Walshe Roussel, L. 2013. The Changing Donor Landscape in Nicaragua: Rising competition enhances ownership and forsters cooperation. *Journal of International Development* 25 (6):802–818.

Wan, C.-D., and M. Sirat. 2018. Internationalisation of the Malaysian Higher Education System Through the Prism of South-South Cooperation. *International Journal of African Higher Education* 4 (2):79–90.

Wells, M. 2011. 'You'll be Fired If You Refuse': Labor Abuses in Zambia's Chinese State-owned Copper Mines. New York: Human Rights Watch.

Whitfield, L. ed. 2008. *The Politics of Aid. African Strategies for Dealing with Donors*. Oxford: Oxford University Press.

Willis, K. 2016. Viewpoint: International development planning and the Sustainable Development Goals (SDGs). *International Development Planning Review* 38 (2):105–111.

Woods, N. 2008. Whose aid? Whose influence? China, emerging donors and the silent revolution in development assistance. *International Affairs* 84 (6):1205–1221.

Woolfrey, S. 2013. The IBSA Dialogue Forum ten years on: Examining IBSA cooperation on trade. Stellenbosch: TRALAC Trade Law Center.

World Bank. 2018. *Indicators*. The World Bank, ND 2018 [cited 18 June 2018]. Available from https://data.worldbank.org/indicator?tab=featured

Xiaofang, S. 2015. Private Chinese Investment in Africa: Myths and Realities. *Development Policy Review* 33 (1):83–106.

Young, K. E. 2017. A New Politics of GCC Economic Statecraft: The Case of UAE Aid and Financial Intervention in Egypt. *Journal of Arabian Studies* 7 (1):113–136.

Zhang, D., and H. Shivakumar. 2017. Dragon versus Elephant: A Comparative Study of Chinese and Indian Aid in the Pacific. *Asia & the Pacific Policy Studies* 4 (2):260–271.

Zhou, H. 2012. China's evolving aid landscape: crossing the river by feeling the stones. In *Development Cooperation and Emerging Powers*, eds. S. Chaturvedi, T. Fues and E. Sidiropoulos, 134–168. London: Zed Books.

Zimmerman, F., and K. Smith. 2011. More actors, more money, more ideas for international development co-operation. *Journal of International Development* 23:722–728.

Index

Entries in *italics* denote figures; entries in **bold** denote tables.

ABC (Brazilian Cooperation Agency) 97
Abu Dhabi Fund for Development 67
accountability 17, 115, 118
Addis Ababa 10, 108–9
Addis Ababa Action Plan 10
Africa: appearances in titles of publications
    20; and BRIC(S) 51–2; Chinese
    migration to 78–82, *81*; Chinese traders
    in 140, 142–3; Chinese-owned factories
    in 81–2, 137–8, 142, 144; data on
    Chinese aid projects in **22**, 24; flows to
    161; higher education in 85; national
    economies of 139–40; SEZs in 77;
    South-South investment in 74, *75*; year
    of 1, 12
African Renaissance Fund 102
African Union, and Turkey 12
Afro-Arab summits 12
AGOA (African Growth and Opportunity
    Act) 71–2
aid: Chinese use of term 129; data on 21–3;
    definition of 16, 21, 63; political nature
    of 62, 65; purpose of 15, 17, 66, 130;
    South-South and North-South 62–6; and
    trade 17; *see also* development finance;
    South-South aid
aid effectiveness 14, 115, 117–19
aid fatigue 114, 120, 130
AidData 21–2, 24, 89, 162
ALBA-TCP (Bolivarian Alliance for the
    Americas - People's Trade Agreement))
    55–7, 60, 148, 158
Algeria 77
Alliance Française 86

Angola: Chinese involvement in 41, 145;
    Cuban engagement in 32, 40–1
Antigua and Barbuda 56
APSA (Summit of South American-Arab
    Countries) 12–13
Arab-Africa Initiative Conference 26n3
Argentina, and development economics 37
Asian financial crisis of 1997 47
authoritarian regimes 18, 59, 150, 153

Bandung Conference 25, 28–9, 32, 34, 45,
    157
Bangladesh 49, 99
Benin 69
bilateral aid 64–5, 90n3
Bolivia 56
Brasília Declaration 52–3
Bräutigam, Deborah 21–2
Brazil: aid system of 97; and BRIC(S)
    49–50, 53; civil society in 102; and
    development economics 37; and IBSA
    52; and Lusophone Africa 8, 95;
    manufacturing in 45; and ProSAVANA
    151; technical cooperation *98*
Bretton Woods institutions 14, 39, 123
BRIC(S): and Africa 139; divisions among 60;
    and IBSA 52, 158; investment in Africa 74;
    leaders' summits 50–1; and NAM/NIEO
    59; and New Development Bank 87–8; use
    of term 49–50; website of 60
British Council 86
Buenos Aires Plan of Action 25; and
    Nairobi protocol 58; and TDC 117
Busan, fourth HLF in 115, 117–18, 130

Caldor, Nicholas 37
Cambodia 28, 146
CARI (China Africa Research Initiative) 22, 25
Castro, Fidel 40, *41*
Centre for Indian Studies in Africa 85
the *Cerrado* 151
Charter of Economic Rights and Duties of States 37
China: African traders in 94, 102–5, *104*, 161; appearances in titles of publications *20*; in Cold War 15; and debt sustainability 122–3; economic growth of 48–9, 138–9; and educational SSC 84–5; and Egypt 8; foreign policy of 32–4; health aid from 149; and new imperialism 14; oil investments in Sudan 109–11; and Pacific Islands 62–3; political criticism of 143–4; SEZs established by 76–7; and TAZARA railway 33–4; trade with Global South 134; voting power in World Bank 124–5
China-Africa relations: bilateral 32; different actors involved in 94–5; and FOCAC 1, 9–10; and Global North 13; global warming in 126; and human rights 149–50; impact on Africa of 133, 138–40, 161; impact on West of 159; information on 18; and migration 81; occurrence on Google Books 19, *20*; *see also* Chinese aid, to Africa
China-DAC Study Group 115, 131
Chinese aid: to Africa 62, 126–9, **127–8**; convergence with Global North 129–30, 159; data on 16, 21–2; entities involved in 93, 95–7, *96*; entities involved in 112–13; principles of 35, 150; purposes of 66; supporting authoritarian regimes 150, 153
Chinese firms: incentives for internationalisation 161; private 101; state-owned 97
Chinese language 85–6; *see also* Confucius Institute

Chocolate City 103, 141
civil society: aid from Global North and 66; in Global South 69; and IBSA 53; in SSC 101–2
climate change 121, 126
Cold War 15, 25, 28–9, 34–5, 39–41
conditionalities: absence of 14, 17, 114, 121; in North-South aid 65–6, 114
Confucius Institute 84–6
Contingent Reserve Arrangement 59, 87, 90n10
copper 33–4, 123, 148
Corea, Gemani 38, 43n2
Corporate Social Responsibility 163
corruption 109–11
Cuba: aid from 15; and ALBA-TCP 56; presence in Africa 32, 40–1

DAC (Development Assistance Committee): definitions of aid 21; development policy statements from members **120**; and emerging donors 7–9, 15, 90n2, 114–15, 117, 120–1, 157, 159; principles of 15–17, 63, 65–6; UAE and 67–8
DCF (Development Cooperation Forum) 116, 131
debt crisis 15–17, 39, 42, 66, 122
debt relief 13, 16, 64–5, 97, 119
debt sustainability 122–3, 147
decolonisation 28, 37
deindustrialisation 42, 139
Delhi Declaration 11
democracy 8, 17, 53, 62
Denmark, development policy of 121
developing countries: China on 34; disagreements between 42; use of term 6
developing world 5–6, 133
development: depoliticisation of 157; Southern and Northern actors in 16
development age 14–15
development aid *see* ODA
development assistance 16–17

development cooperation: international 14–15, 66, 114, 116; 'as it was' 3–4, 8, 14, 129; Southernisation of 120, 126, 129–30
development economics, emergence of 36–7
development effectiveness paradigm 119
development finance: conditionality of 17; data on 22, 162; differences between South-South and North-South 63–5, **64**; private firms in 99–100; and South-South cooperation 13, 35, 158; terms used for 16
development sustainability 122, 147
division of labour, global 32, 38, 45–6
Dominica 56
donor meetings, common 64–5
DRC (Democratic Republic of the Congo) 122–3

economic development: and growth 37; and NAM 29
economic growth: aid focus on 66, 119, 121; and development 37; large companies and 94; rapid 1, 7, 46–9, 145; and trade 70; in Venezuela 6
Ecuador 56, 69
Education, in Global South 83–5; *see also* higher education
Egypt: manufacturing in 49; SEZs in 77; South-South development cooperation with 8; UAE aid to 67–8
Eight Principles of Chinese Assistance 34–5
EMBRAPA (Brazilian Agricultural Research Cooperation) 97
emerging donors: and Cambodia 146; Global North reactions to 14–18, 119, 121; and TDC 116; use of term 7–9
emerging South 6, **7**, 42, 69, 157
EMNC (emerging multinational corporations) 76
employment conditions, abusive *see* labour rights

Erdoğan, Recep Tayyip 12
Ethiopia: Chinese migrants to 81; Cuban presence in 40; energy sector in 147; SEZs in 77; shoe and leather industry 94, 102, 107–9, 135, 161
EU (European Union), and emerging donors 8
ExIm Bank of China 96–7, 123
export credits 16–17, 64

FDI (Foreign Direct Investment): data on 23; developmental effects of 135–6; forms of 90n5; from Global South 74, 160; *see also* South-South investment
Fiji 62–3
Financial Tracking Service 69, 89
First/Developed World 5
flying geese concept 138
FNLA (National Liberation Front of Angola) 41
FOCAC (Forum on China Africa Cooperation): and Chinese aid **127–8**; declarations from 10; and education 85; launch of 9; program of 2; reactions from Global North 13–18; reactions from rest of South 10–13; website of 25
FOCAC I (Beijing, 2000) 9
FOCAC II (Addis Ababa, 2003) 10
FOCAC III (Beijing, 2006) 1, *2*, 13, 18, 126, 157, 162
FOCAC IV 126
FOCAC V 126
FOCAC VI 126
food security 57, 151
footwear industry 107–9
Forum for Indian Development Cooperation 101
Foshan 103–5
Furtado, Carlos 36

G8 60n2; and outreach 5 52
G20 53, 60–1n3
G77 25, 36, 42–3, 116

Gates, Bill and Melinda 87
GATT Uruguay Round 42
Ghana: employment conditions in 150; imports of motorbikes 133–4; trade in Chinese goods in 140, *141*; traders in China from 105
global financial crisis of 2008: and aid discourse in North 120; and ALBA-TCP 57; and BRIC(S) 49–50, 87; and FDI from Global South 74; and IFIs 123; impact on Africa of 144
global governance: and BRIC(S) 50; economic and financial 50; and Global South 4; SSC in 87–9, 137; structures of 132, 157
Global North: acknowledgement of China and India 49; and Global South 4–5, 38 (*see also* North-South convergence); and humanitarianism 68; reaction to SSC 114–16, 118; and SSC 4; support for NIEO from 43n2
Global South: benefits from trade 160; border control in 78; and BRIC(S) 51–2; changes in 130; demand for manufactures 71; flows within 65, 158–9; heterogeneity of 6–7, 41–2, 46, 137, 157; humanitarian aid from 69; investment from 58, 76; manufacturing in 36, 39, 47–9; relocation of industries within 138–9; and SSC 3; use of term 4–6; waves of regionalism in 55; *see also* emerging South; rest of the South
Global South Studies Center 26
Global Value Chain approach 93, 160
globalisation: ALBA's critique of 56; and FOCAC 2; and Global South 5
Goethe Institute 86
Goldman Sachs 49–50
good governance 2, 17, 150, 159
grannational projects 57
greenfield investments 74, 90, 136
grey literature 20
Guangdong 140–1

Guangzhou 103–7, *104*, 140
Guinea, traders in China from 105
Gulf states 21, 32, 67, 90n4

Haiti 69
Hanban (Office of Chinese Language Council International) 85–6
HDI (Human Development Index) 4
health sector: in Cambodia 146; and DAC 120; and IBSA 55; in Madagascar 149; under ALBA 57; under FOCAC 9
Helsinki Package of Tied Aid Disciplines 17
heterodox economics 36
higher education: African students in China 18; internationalisation in Malaysia 83–4; SSC in 40, 85
HIPC (Highly Indebted Poor Countries Initiative) 17
HLFs (High-Level Fora) 115, 117–18, 131
Hong Kong 45, 47, 71, 73, 80
Hu Jintao 1
human rights: and Bandung conference 29; Global North focus on 2, 17, 86; IBSA and 53; NGOs and 102; and SSC 149–50, 162
Human Rights Watch 150–1
humanitarianism 63, 67–70
hydropower projects 112, 147–8

IBSA (India-Brazil-South Africa Dialogue Forum) 49, 52–5, 59, 137, 158; trade within **54**
ICT (information and communications technology) 49, 84
IFIs (international financial institutions): and Angola 145; and BRIC(S) 50, 87; and debt sustainability 122, 147; governance and human rights agenda 2; and NICs 47; projects financed by 94; reactions to emerging donors 121–3, 125–6, 130
IMF (International Monetary Fund): and Angola 145; bailout of 87; and Contingent Reserve Arrangement 90n10;

data from 23; and debt sustainability 122–3; establishment of 14, 28; president of 50, 60n1; voice reforms in 123, **125**
import substitution 32, 36–7
India: actors involved in aid 93, 101; and Africa 10–11, 18; aid to Pacific Islands 62–3; bilateral South-South Cooperation by 32; and BRIC(S) 53; development assistance from 16, 21, 35, 62; economic growth in 48–9; and educational SSC 84–5; and Egypt 8; export of capital equipment 133; female peacekeepers from 99; multinational corporations from 32; and new imperialism 14; oil investments in Sudan 18, 109–11; voting power in World Bank 125
Indian diaspora 62–3
Indonesia 28–9, 47, 75
industrialisation 128, 138–9, *140*
Institute of Development Studies 133
internationalisation: of firms from Global South 35, 45, 72, 76, 79, 82, 93, 97, 161; of higher education 83–4
Israel 73
ITEC (Indian Technical and Economic Co-operation) 84
Ivory Coast 46

Japan: aid to Cambodia 146; and DAC 9; and ProSAVANA 151
jewellery, Chinese 140, *141*
Jiang Zemin 9
Jordan 46, 73

Kabila, Laurent 122
Kaname, Akamatsu 138
Kaunda, Kenneth 33
Kenya 49, 69, 105, 133, 135
knowledge transfers 146, 149
Kuwait 35, 67, 122
Kuwait Fund for Arabic Economic Development 35
Kuznets, Simon 37

labour, low-wage 38, 48, 73, 77
labour rights 143, 150–3, 162–3
Lagarde, Christine 60n1
Latin America: China in 163; and development economics 36–7; lack of English-language literature on 24; multinational corporations from 32; regional integration in 6; re-politicisation of 57
League of Arab States 12
Lehman Brothers 50
Lesotho 71, 152
Lewis, Arthur 37
Li Zhaoxing 1
Liberia, female Indian peacekeepers in 99
Lin, Justin Yifu 138
loans, concessional 16, 33–4, 64, 146
Lula da Silva 95
Lusaka 82, 105–7
Lusophone Africa 95

Madagascar 71, 149
Malaysia 47; FDI in Vietnam 75–6; higher education in 83–4
Mali 28, 105
Mandarin *see* Chinese language
Manning, Richard 14, 114
manufacturing: access for South to Northern markets 36, 39; labour-intensive 76, 138–9; light 47, 49, 77; in NICs 45, **46**; Northern exports to South 38; relocation of 139–41; South-South trade in 31
Mao Zedong 79
Maritime silk road 126
Mauritius 47, 71, 77, 80
Mawdsley, Emma 126, 129
MDGs (Millennium Development Goals) 26n3, 115, 117, 119, 147; website for 131
Merkato footwear cluster *108*
MFA (Multi-Fibre Agreement) 71–3, 135
migration 77–83; global stock **78**; *see also* South-South migration

Ministry of Commerce (China) 16, 97, 113, 129
Modi, Narendra 101
Mozambique 33, 40, 151
MPLA (Movement for the Liberation of Angola) 40–1
multilateral aid 64–5
multinational corporations: from Global South 32, 47, 74; power to regulate 39; and SSC 3–4
mutual benefit: Chinese principle of 9–10, 34–5; DAC's adoption of 17, 66, 120–1; Global North's reaction to 4, 114
Mwanavasa, Levy 82
Myrdal, Gunnar 37

Naím, Moises 14
Nairobi protocol 58
NAM (Non-Aligned Movement): and BRICS/IBSA 59; establishment of 29; and UNDP 30
Namibia 18
Nana-Benzes 142–3
National Congress Party (Sudan) 110–11
Nehru, Jawaharlal 28
neoliberalism 5, 24, 39, 57, 145, 157
the Netherlands 9, 121
Network of Southern Think Tanks 60
New Development Bank 59, 87, 88, 90n10
NGOs see civil society
Nicaragua 18, 56, 147–8
NICs (Newly Industrialised Countries) 45–8, 75
NIEO (New International Economic Order) 24–5, 43n1; and ALBA-TCP 57; and BRICS/IBSA 59; call for 29–30; proponents of 43–4n2; UN ratification of 37–9
Nigeria: APSA summit in 13; motorbikes made in 133–4; SEZs in 77; traders in China from 105
non-interference 150

North-North trade 70
North-South aid 63–6
North-South convergence 126, 158–9
North-South Cooperation 58, 117
North-South investment 74
Nyerere, Julius 33

ODA (overseas development assistance): and DAC 63–6; India and 11; terms used in 5–6; use of term 16–17
OECD (Organisation for Economic Cooperation and Development): and DAC 7–8; data from 21, 23, 89
Official Development Effort 66
offshoring 23, 139–40
oil crises 30–2, 39–42, 46
oil investments 94, 109–11
Olympic Games, Beijing 2008 1
One Belt One Road 129
OOF (other official flows) 63–4
OPEC (Organisation of Petroleum Exporting Countries) 15, 35, 67, 83, 159
Organisation for African Unity 9

Pacific Islands 62–3
Pakistan 73, 99
peaceful co-existence, five principles of 34
peacekeeping forces 99, 113n2
Pérez-Guerrero Trust Fund 30
petrodollars 39, 42
Philippines 47, 69
Point Four Programme 14–15
policy space 145, 147–8, 153, 162
poverty reduction 9, 119–21, 129, 148
Poverty Reduction Strategy Papers 147
power relations: and Global South 6; SSC's impact on 2–3, 19
Prebisch, Raul 36
preferential trade agreements 53, 73
primary commodities: Chinese demand for 48, 135; and global financial crisis 144; price fluctuations in 42, 148; Southern exports to North 45; South-South trade in 31

ProSAVANA programme 151, 154
public goods, global 4, 62, 121, 130, 159

Qatar 6, 12, 67

Red Cross 68, 100
regional cooperation fora 55–9
rest of the Global South 4, 6, 42, 157;
    Chinese and Indian firms in 150; and
    SSC 19
rogue donors 14, 126, 153

Saint Kitts and Nevis 56
Saint Vincent and the Grenadines 56
sameness, imaginary of 100
Santa Lucia 56
Sanyuanli 103
Sata, Michael 82, 143–4
Saudi Arabia: aid from 67; and
    development finance 35;
    underreporting of 24; voting power
    in World Bank 125
Scott, Guy 144
SDGs (Sustainable Development Goals)
    119–20, 131
Seers, Dudley 37
self-reliance 29–31, 34, 39, 44n2
Senegal 6, 105
SEZs (Special Economic Zones) 76–7
SIDCA (State International Development
    Cooperation Agency) 112–13n1
Singapore 6, 45, 47, 75
Singer, Hans 37
Sirleaf, Ellen Jonson 99
al-Sisi, Abdel Fata 67
SOEs (state-owned enterprises) 16, 81, 93,
    97, 161
soft power 69, 84, 86
Solferino, battle of 68
Somalia 69
South Africa: aid agencies 93; aid data
    from 22; and Angola 41; and BRIC(S)
    51, 53; Chinese migrants to 79, 80;
    civil society in 102; incentives for

internationalisation 161; Indian support
    for education in 85; and textile trade 71
South Centre 60
South Korea: aid to Cambodia 146; and
    DAC 9, 90n2, 159; manufacturing in 45;
    as NIC 47
South Sudan 110
Southern Africa 35, 40
South-North migration 78
South-South aid 62–6, 158
South-South Conference 60
South-South Development Cooperation 1,
    13, 26n2, 34, 58; see also SSC
South-South economic integration 31–2
South-South Experience Exchange Facility
    115
South-South flows see Global South, flows
    within
South-South humanitarianism 68–9
South-South Information Gateway 26
South-South investment 23, 32, 74, 75,
    135, 158, 160, 162
South-South migration 78–83, 90n7, 93,
    137, 160
South-South summits, cross-regional 12
South-South trade 7, 24, 31, 32, **70**, 71–3,
    135, 158–9; economic impacts of 134,
    135
Soviet Union 15, 40–1
Special Drawing Rights 87
Sri Lanka 47
SSC (South-South Collaboration): bilateral
    32–6; and BRIC(S) 52; civil society in
    101–2; competitive and complementary
    impacts of 135, **136**; direct and indirect
    effects of 133–7, 161–2; drivenness
    of 93–5, 111–12; economic effects of
    137–42, 152; in education 84–5; and
    FOCAC 1, 10–11; future research into
    162–3; IBSA and 53–4; international
    governance of 36; long-term impact
    of 107; major insights into 156–62;
    occurrence on Google Books 20;
    political developments towards 29–31;

political effects of 143–8, 152–3; private
   actors in 99–100, 161; reaction from
   Global North to 14, 17, 114, 118, 122,
   130; resurgence of 2–3; social effects
   of 149–53; sources of information on
   18–24, 26n4; temporary collapse of
   39–42; vectors of engagement 97
Strauss Kahn, Dominic 60n1
structural adjustment programmes 17,
   66, 122
structuralist economics 24, 37, 59
Sudan: Chinese migrants to 80–1; oil
   companies in 18, 94, 109–11
Sukarno 28
Swaziland 71

Taiwan: manufacturing in 45; as NIC 47;
   tech companies 47–8; and textile trade
   71, 73
Tanzania: Chinese migrants to 79;
   humanitarian aid from 69; and
   TAZARA railway 33–4; traders in
   China from 105
tariff preferences 10, 36–7
tariffs 24, 71, 76, 107–8, 142, 158
Tata Group 100
TAZARA railway 33–4, 79
TDC (trilateral development cooperation)
   65, 114–17, 131n1, 151
technical cooperation 26, 30–1, 58, 117
textiles and apparel: Chinese migrants
   working in 80; greater Chinese
   manufacturers 72–3; impact of South-
   South trade in 135; trade agreements on
   71–2; wax prints 142–3
Thailand 47, 75, 117
Third World 4–5
trade: data on 23–4, 142; development
   economics on 37; global patterns of 70,
   72; see also South-South trade
Trade Law Centre 89
trade unions 40, 144, 150, 152–3
transparency 4, 17, 115, 129
Truman, Harry 14

Tunisia 28, 47
Turkey, and Africa 11–12

UAE (United Arab Emirates): aid
   agencies 93; aid to Egypt 66–8; and
   development finance 35; incentives
   for internationalisation 161; lack of
   conditionality 122
Uganda 22, 28, 105
UN (United Nations): establishment of 28;
   South-South Cooperation within 30, 37,
   58, 87; and TAZARA railway 33
UN Conference on Trade and Development
   (UNCTAD): data from 23, 89; and G77
   36; and investment 74; and NIEO 38;
   and SSC 58; website of 43
UN Nations Development Program 30, 55
UN Office for South-South Collaboration
   26, 30, 43
UN Security Council 11, 35, 52–3, 59, 150
UN Volunteer Programme 100
UNITA (National Union for the Total
   Independence of Angola) 41
United Kingdom: development policy of
   120–1; promoting English language 86
United States: Cuban conflict with
   40; dollar 34, 50; and South-South
   Cooperation 14; textile imports to 72–3

value addition 45, 70–2
value chains: of cheap jewellery *141*;
   global 91, 140; in textiles and apparel
   72–3; see also Global Value Chain
   approach
Venezuela: aid to Nicaragua 148; and
   ALBA-TCP 56; and authoritarian
   regimes 150, 153; growth trajectories of
   6; underreporting of 24
Vietnam 28, 75–6
Vlisco 142–3
voice reforms 123, **124**, **125**

Washington Consensus 2
wax print fabrics 142–3

Wen Jiabao 1
white supremacy 33, 35, 40
win-win outcomes 13, 94, 120–1, 134, 137–8, 158
Working Group on Technical Cooperation among Developing Countries 30
World Bank: and conditionality 122; data from 23; and emerging donors 115; establishment of 14, 28; and hydropower 147; president of 50, 60n1; and TAZARA railway 33; voice reforms in 123, **124**, **125**
World Food Programme 69
World Investment Report 74
World War II, end of 28
WPAE (Working Party on Aid Effectiveness) 114–15
WTO (World Trade Organisation) 54, 137

Xiaobei 103

Yiwu 103, 105

Zambia: aid from 69; Chinese migrants to 79, 81–3; Confucius Institute in 85–6; mining companies in 74; overseas investments in 139; political opposition to China in 143–5, 150–2; relations with donors 148–9; SEZs in 77; South African goods in 135; and TAZARA railway 33–4; trading with Chinese goods in 105–7
Zambia Consolidated Copper Mine 81
Zenawi, Meles 109
Zhejiang 140
Zhou Enlai 33–4
Zoellick, Robert 60n1